Get Through

MRCS: Anatomy Vivas

This book is dedicated to Lucas.

Get Through
MRCS: Anatomy Vivas

Simon Overstall MBBS MRCS

Plastic Surgery Registrar
The Alfred Hospital
Melbourne, Australia

The ROYAL
SOCIETY *of*
MEDICINE
PRESS *Limited*

© 2006 Royal Society of Medicine Ltd

Published by the Royal Society of Medicine Press Ltd
1 Wimpole Street, London W1G 0AE, UK
Tel: +44 (0)20 7290 2921
Fax: +44 (0)20 7290 2929
E-mail: publishing@rsm.ac.uk
Website: www.rsmpress.co.uk

British Library Cataloguing in Publication Data
A catalogue record for this book is available from the British Library

ISBN 1 85315 684 1

Distribution in Europe and Rest of World:

Marston Book Services Ltd
PO Box 269
Abingdon
Oxon OX14 4YN, UK
Tel: +44 (0)1235 465500
Fax: +44 (0)1235 465666
E-mail: direct.order@marston.co.uk

Distribution in the USA and Canada:

Royal Society of Medicine Press Ltd
c/o Book Masters Inc
30 Amberwood Parkway
Ashland, OH 44805, USA
Tel: +1 800 247 6553 / +1 800 266 5564
Fax: +1 419 281 6883
E-mail: order@bookmasters.com

Distribution in Australia and New Zealand:

Elsevier Australia
30-52 Smidmore Street
Marrikville NSW 2204, Australia
Tel: +61 2 9517 8999
Fax: +61 2 9517 2249
E-mail: service@elsevier.com.au

Typeset by Phoenix Photosetting, Chatham, Kent
Printed by Bell & Bain Ltd., Glasgow

Contents

Acknowledgements

Many thanks to Dr Norman Eisenberg and Dr Chris Briggs at the University of Melbourne for the use of the Anatomedia images. Also thanks to Dr Gerry Ahern and the 'Dartos Contractors' of Monash University for the use of their anatomical images. Thanks to Dixon Woon and Natalie Zantuck for proofreading.

A special thanks to Freya for all of her support during the writing of this book.

Introduction

Anatomy is one of the toughest subjects for medical students and surgical trainees. It requires a good memory for detail and the ability to think in three dimensions. There is (unfortunately) no shortcut to learning the subject. Read through the books and then consolidate your knowledge using dissected cadavers, plastic or computer models, or time spent in the operating theatre. Ask questions whenever assisting in theatre. Most surgeons are only too keen to show off their hard-earned knowledge of anatomy.

However, anatomy is more than just rote learning of tables. Understand the basic layout of the body, the concept of fascia and compartments, and visualize the three-dimensional model in your mind's eye. There are a few simple rules that may help you deduce many surprise viva questions.

Some people find mnemonics a useful way to learn anatomy. My advice would be to use mnemonics only in conjunction with an understanding and a mental picture of the region. It is very easy to remember the mnemonic, but to forget what the dirty rhyme or acronym actually stands for. I have included some of the cleaner mnemonics for those of you who have not heard them before.

This book is designed to be used as a self-testing tool in preparation for the anatomy viva voce. The list of topics discussed here is based on previous viva questions remembered by candidates sitting the MRCS exam over the past few years. By no means is the list exhaustive. Certain topics crop up in the exams time and time again. I have tried to represent the prevalence of these popular questions in this book. For instance, the abdomen is a much more popular region for viva questions than the back.

The best time to use this book is in the weeks running up to the exam, after having read through a good anatomy textbook and atlas. Close your textbooks and then ask yourself the questions in the front of this book. Better still, find someone else to ask you the questions. Pretend you are in the exam and try to verbalize the answers in full for each question before checking the answers. You'll be surprised at how difficult it is to accurately and concisely explain a topic you thought you knew well. It is much better to practise this technique now before the exam. Don't worry if you can't recite all the answers as listed in this book. These are designed as model answers. You are unlikely to fail for missing some of the finer details.

This book is useful for identifying gaps in your knowledge. Go and read the relevant chapter in the textbook if you don't feel you answered a particular question well enough.

There are six sections in the MRCS viva voce. These are divided into three separate exams:

- Applied Surgical Anatomy/Operative Surgery
- Applied Physiology/Critical Care
- Applied Surgical Pathology/Principles of Surgery.

Each exam lasts for 20 minutes, divided into two 10-minute segments.

During the anatomy section the examiners will often use props. These may be bones, CT scans, MRI scans, specimen pots, live models, photographs or cadavers.

There are two examiners for each section; one surgeon and an examiner of the basic sciences. One will ask the questions whilst the other scores the candidate and then they swap around. You will usually be asked questions on three or more separate topics.

You will probably be asked a variety of questions ranging in difficulty. The examiners will usually start with easier questions and then progress to greater depth if you answer them well. However, don't read too much into this. Just because you are asked several easy questions doesn't mean you answered the previous one badly.

There are some general tips that are useful for any viva voce:

- Always ask to have the question repeated or rephrased if you didn't hear or didn't understand the first time.
- Stop and think for a few seconds before shouting out the answer. This will give you a chance to organize your thoughts.
- Try to present your answer in a clear, logical way. If possible, break your answer up into headings or topics, e.g. What are the branches of the axillary artery? 'The axillary artery is divided into three parts according to the relationship to the pectoralis minor muscle. The first part has one branch, the second part has two branches and the third part has three branches. They are as follows...' A snappy, concise answer like this will make it clear to the examiner that you know your stuff, he'll tick his box and move on to the next question. A long-winded list of the branches that have to be dragged out in a random order will annoy the examiners and lead to lower scores.
- Always look the examiners in the eye. You will often be handed a bone or some other prop to lead the questions. It is a bad idea (but often happens) to mumble the answer whilst staring at the object. Look at, and speak to, the examiners; use the prop only to demonstrate certain points in your answer.
- Expect the unexpected. There will always be a question you won't know the answer to. If this happens it is reasonable to make an informed guess. However, if you really don't know then say so. The examiners will move on to another topic (hopefully one that you will know) and you'll have another chance to pick up points.
- Put the bad ones behind you. Don't dwell on the question you didn't know or answered badly. Try not to get too flustered. When that question has finished you should move on and concentrate on the next subject.
- When shown a cadaver or wet specimen, take your time to orientate yourself. Find a point of reference that you recognize and work your way round from this. For example, when shown the brachial plexus, look for the 'M' shape that is formed by the confluence of the branches from the medial and lateral cords that form the median nerve. You should then be able to work your way back from this landmark and deduce the rest.
- Be concise with your answers. Avoid long waffle if you can answer the question in a few clear sentences. This will impress the examiners and allow you more time to gain points in the next question.

Practice Viva Questions

1. Upper limb

1.1   Draw the brachial plexus and label the branches.

1.2   Which muscles do the branches of the posterior cord of the brachial plexus supply?

1.3   Which muscle does the long thoracic nerve supply and how would you test it?

1.4   How many branches arise from the divisions of the brachial plexus?

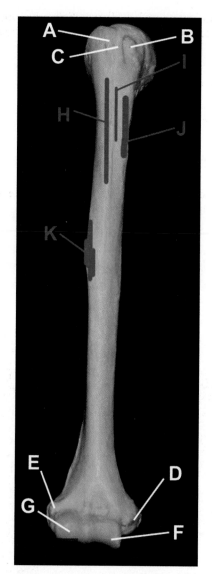

Reproduced with permission of Dartos Contractors (sca78548@bigpond.com.au).

1.5

    a Identify this bone.
    b Identify the bony points A to G.
    c Which muscles attach at points H to K?

1.6   Describe the arrangement of the muscles in the flexor compartment of the forearm.

1.7   What is the nerve supply to the forearm flexor compartment muscles?

1.8   What are the actions of the flexor digitorum profundus and flexor digitorum superficialis muscles? How would you test them?

1.9   What is the motor nerve supply to the hand?

1.10   What do the lumbricals do?

1.11   What do the interossei do?

1.12   How many compartments are there in the extensor retinaculum and what runs through each one?

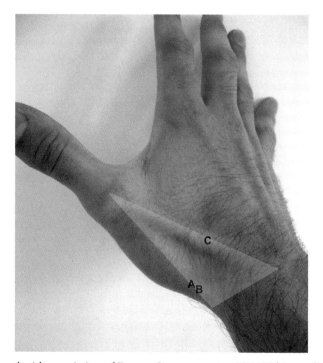

Reproduced with permission of Dartos Contractors (sca78548@bigpond.com.au).

1.13

   a What is the yellow shaded region on the dorsum of the hand known as?
   b What are the boundaries of this region? Which tendons are labelled A to C?
   c What are the contents of this anatomical region?

1.14   Which muscles make up the shoulder rotator cuff?

1.15 What are the boundaries of the axilla?

1.16 What are the contents of the axilla?

1.17 Name the branches of the axillary artery.

1.18 What is the clavipectoral fascia and which structures pierce it?

1.19 What is the quadrangular space and which structures pass through it?

1.20 Which nerves can be damaged by a fracture of the humerus?

Screw the lawyers, save a patient

STl SAP

Sup. thoracic
Thoraco-acromial
Lat. thoracic
Subscapular
Ant. humeral circumflex
Post. humeral circumflex

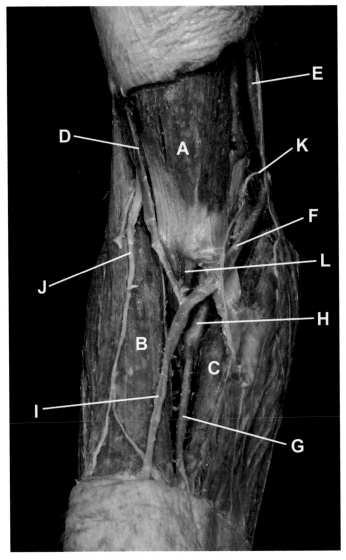

Reproduced with permission of Dartos Contractors (sca78548@bigpond.com.au).

1.21

    a Identify the muscles labelled A to C.
    b Identify the other structures labelled D to L.

1.22   What are the boundaries of the cubital fossa?

1.23   What are the contents of the cubital fossa?

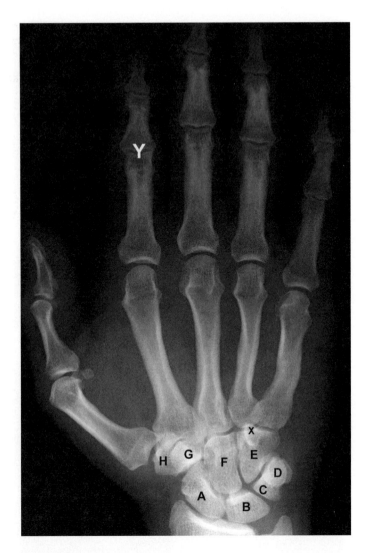

1.24

a Name the carpal bones labelled A to H.
b What is marked by the X?
c Which joint is marked by the Y?

1.25 Which of the carpal bones provide attachment for the flexor retinaculum and which structures pass through the carpal tunnel?

1.26 Which structures are cut through during an open carpal tunnel release and which structures may be damaged during this procedure?

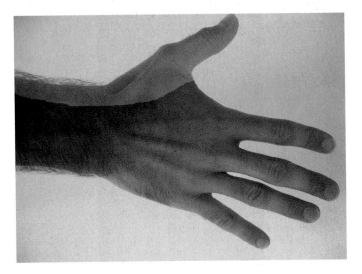

1.27 Which dermatome is this?

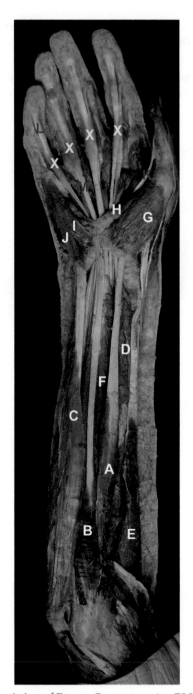

Reproduced with permission of Dartos Contractors (sca78548@bigpond.com.au).

1.28

   a Identify the muscles labelled A to J.
   b Identify the tendons labelled K to M.
   c What happens to the tendons at the points marked X?

1.29 What prevents the flexor tendons from bow stringing? Describe the arrangement of these structures.

1.30 Where does the musculocutaneous nerve arise from?

1.31 Which muscles does it supply?

1.32 Does this nerve have a sensory component?

2. Lower limb

2.1   How many compartments are there in the leg?

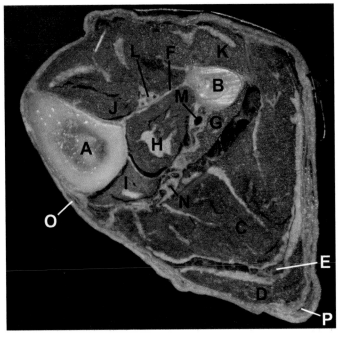

Reproduced with permission of Dartos Contractors (sca78548@bigpond.com.au).

2.2

    a What are the structures labelled A to K?
    b Identify the arteries labelled L to N.
    c Name the superficial structures labelled O and P?
    d Which nerves run next to the structures labelled O and P?
    e Which nerves run next to the arteries labelled L and N?

2.3   Which muscles lie in the anterior compartment of the leg?

2.4   What happens if the nerve to the anterior compartment of the leg is damaged?

2.5   What are the boundaries of the femoral triangle?

2.6   What are the contents of the femoral triangle?

2.7   What is the femoral sheath?

2.8   What is the femoral canal?

2.9   What is the femoral ring and what are its boundaries? What is its significance to the surgeon?

2.10   Where would you palpate the foot pulses?

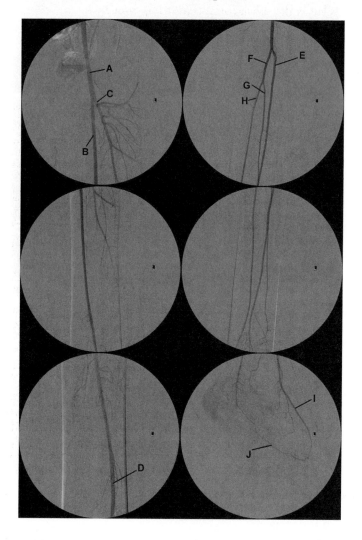

2.11

a What investigation is this?
b Identify the vessels labelled A to J.

2.12 Describe the vascular supply to the leg.

2.13 Describe the path of the long saphenous vein.

2.14 Which nerve runs in close proximity to the long saphenous vein? What is the relevance of this to the vascular surgeon?

2.15 Where does this nerve arise?

2.16 What does this nerve supply?

2.17 What is the subsartorial (Hunter's) canal?

2.18 What are its boundaries?

2.19 What runs through it?

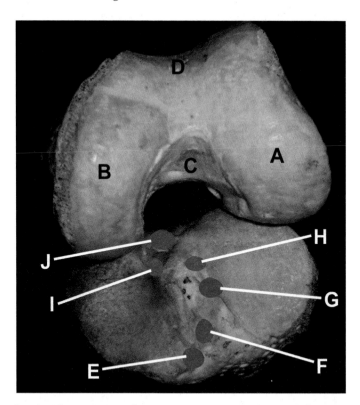

Reproduced with permission of Dartos Contractors (sca78548@bigpond.com.au).

2.20

    a Which joint is this and which side of the body is it from?
    b Name points A to D.
    c What attaches to points E to J?

2.21 What do the cruciate ligaments do? How would you test them?

2.22 Which of the cruciates is the stronger?

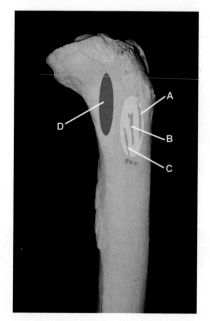

Reproduced with permission of Dartos Contractors (sca78548@bigpond.com.au).

2.23

    a Which bone is this?
    b Name the area highlighted in yellow.
    c What attaches to points A to C?
    d What attaches to point D?

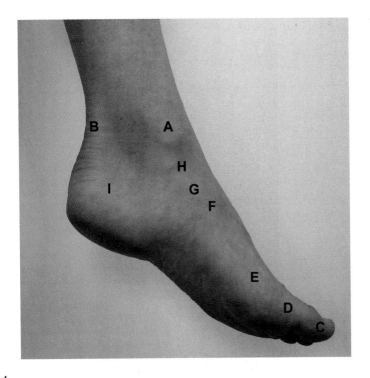

2.24

  a Name the bony prominence labelled A.
  b Which structures pass behind (posterior to) this landmark?
  c Name the structure palpable at point B.
  d Which muscles contribute to this structure?
  e What are the nerve supplies to these muscles?
  f Which bones can be palpated at points C to I.

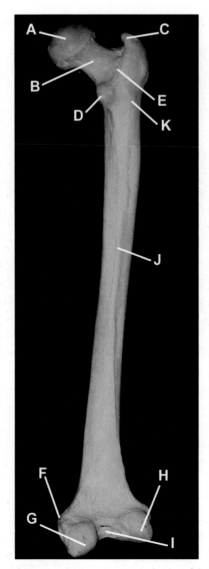

Reproduced with permission of Dartos Contractors (sca78548@bigpond.com.au).

2.25

    a Which bone is this and which side is it from?
    b Name the parts labelled A to K.

2.26 Describe the blood supply to the head of the femur.

2.27 What is the significance of this?

2.28 Name the muscles in the anterior compartment of the thigh.

2.29   What is the nerve to the medial compartment of the thigh?

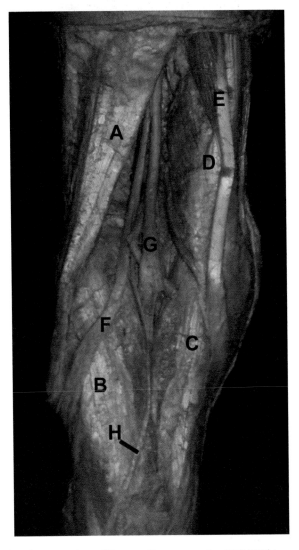

Reproduced with permission of Dartos Contractors (sca78548@bigpond.com.au).

2.30

    a Identify the structures labelled A to H.
    b What does structure F divide into?

2.31   What are the boundaries of the popliteal fossa?

2.32   Which structures are in the popliteal fossa?

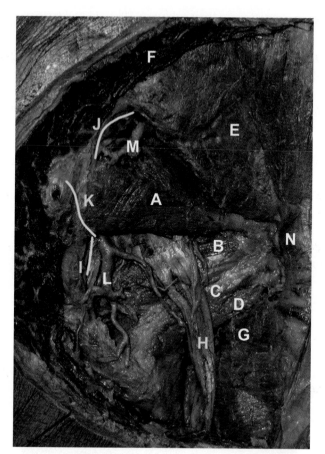

Reproduced with permission of Dartos Contractors (sca78548@bigpond.com.au).

2.33

a This is the posterior aspect of the gluteal region. Which side is it from?
b Identify the muscles labelled A to G.
c Which nerve exits the pelvis above muscle A?
d Which muscles does this nerve supply?
e What do these muscles do?
f How would damage to this nerve manifest itself?
g Identify the nerves labelled H to K.
h Identify the arteries labelled L and M.
i From which artery do these two vessels originate?
j Through which foramen do these two arteries exit the pelvis?
k Which bony prominence is marked by N and which muscles insert here?

2.34 Which structures are cut through during the posterior approach to the hip joint?

2.35   Which nerves are at risk in this approach?

2.36   Describe the muscle layers of the sole of the foot. Which other structures run in these layers?

2.37   Which cutaneous sensory nerve supplies sensation to the outside of the thigh?

2.38   Which dermatome is this?

2.39   What are the surface markings of this nerve (as used to perform a nerve block)?

2.40   What is the sensory nerve supply to the first web space of the foot?

2.41   What are the nerve roots of the sciatic nerve?

2.42   What are the terminal branches of the sciatic nerve?

2.43   How would you test the sensation and motor function of the L5 nerve root?

3. Head and neck

3.1 Describe the arrangement of the deep fascia of the neck.

3.2 What are the boundaries of the posterior triangle of the neck?

3.3 Describe/demonstrate the actions of the sternocleidomastoid and trapezius?

3.4 Which structures are found in the posterior triangle of the neck?

3.5 What would be the effect of cutting the spinal accessory nerve?

3.6 List the layers you would cut through during a tracheostomy tube insertion.

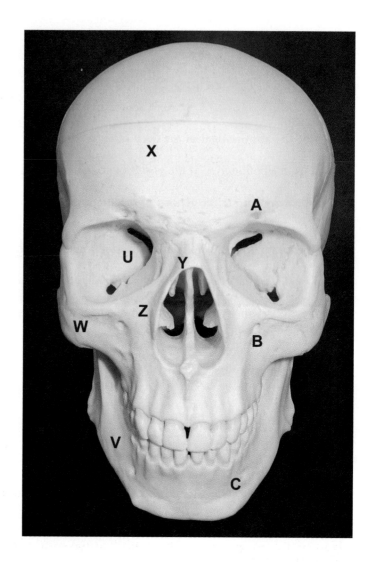

3.7

   a Name the foramina labelled A to C. Which structures pass through them?

   b Name the bones labelled U to Z.

3.8   How many bones does the sphenoid articulate with and what are they?

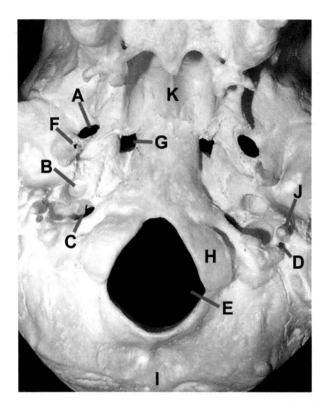

3.9

   a Name these foramina. What structures pass through them?
   b Identify points H to K.

3.10   What is the path of the facial nerve? What are its branches and which structures do they supply?

3.11   How would you quickly test the motor function of the divisions of the facial nerve?

3.12   What is the sensory nerve supply to the ear?

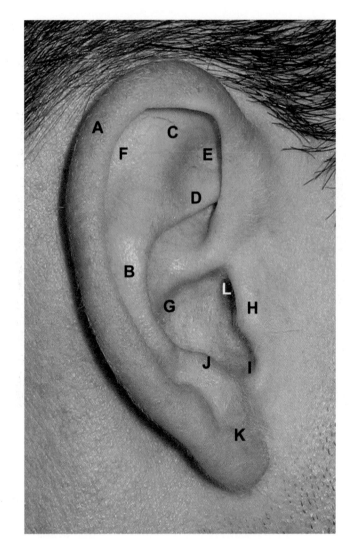

3.13　Identify the different parts of the ear labelled A to L.

3.14 Describe the arrangement of the venous drainage of the brain.

3.15 What are the cavernous sinuses, what passes through them and what is their clinical relevance?

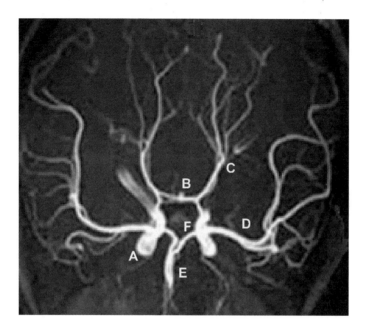

3.16

    a Identify the vessels labelled A to F.
    b Name this anastomosis of vessels.

3.17 What are the surface markings of the cervical vertebrae levels?

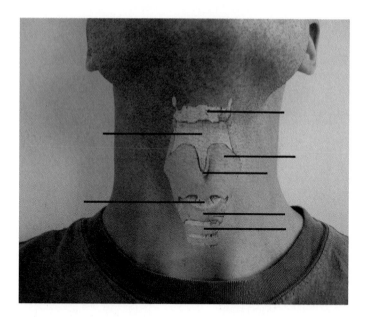

3.18 Surface mark the following parts of the larynx:

- Hyoid bone
- Thyroid cartilage
- Cricoid cartilage
- First tracheal ring
- Thyroid membrane
- Laryngeal prominence
- Cricothyroid ligament

3.19 What are the branches of the external carotid artery?

3.20 What are the branches of the internal carotid artery?

3.21 Which cervical vertebral level corresponds to the bifurcation of the common carotid artery?

3.22 What are the branches of the maxillary artery?

3.23 How many cranial nerves are there and what are their names?

3.24 What are the divisions of the fifth cranial nerve?

3.25 What do they supply sensation to?

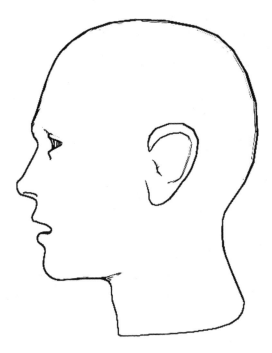

3.26 Draw the sensory distribution of the divisions of the fifth cranial nerve on this face.

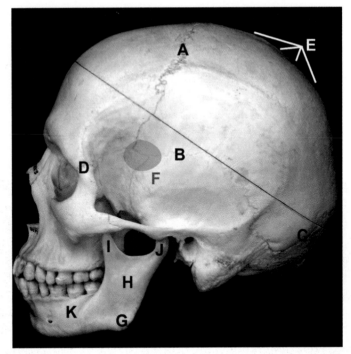

Reproduced with permission of Dartos Contractors (sca78548@bigpond.com.au).

3.27

    a Name the sutures labelled A to E.
    b Name the red shaded area F?
    c What runs behind it?
    d What is the clinical significance of this?
    e Identify the parts of the mandible labelled G to K.

3.28 What are the surface markings of the parotid gland?

3.29 What are the surface markings of the parotid (Stenson's) duct?

3.30 What are the layers of the scalp?

3.31 What is the blood supply to the scalp?

3.32 Where is the standard incision made for access to the submandibular gland? Why is it made here?

3.33 In what anatomical triangle does the submandibular gland lie?

3.34 What nerves may be damaged during a submandibular gland excision?

3.35 Which vessels are usually encountered during the approach to the submandibular gland?

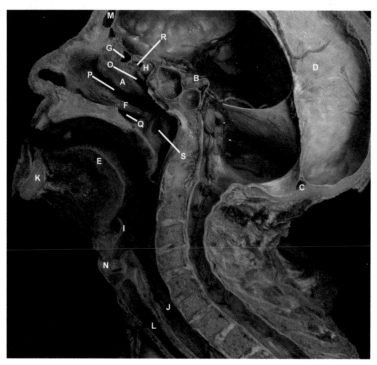

Reproduced with permission of Dartos Contractors (sca78548@bigpond.com.au).

3.36

    a Name the structures labelled A to N.
    b Identify the structures labelled O to S. What do they communicate with?

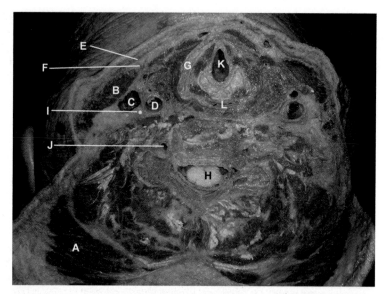

Reproduced with permission of Dartos Contractors (sca78548@bigpond.com.au).

3.37 What structures run through the superior orbital fissure?

3.38 Describe the arrangement of the extraocular muscles, their actions and nerve supply.

3.39

    a Identify the structures labelled A to L.
    b Which vertebral level is this cross-section?

3.40 Which vertebral level does the isthmus of the thyroid gland correspond to?

3.41 Describe the relationship of the thyroid gland to the layers of the cervical fascia.

3.42 Describe the blood supply to the thyroid gland.

3.43 Which structures are at risk of damage during ligation of these vessels during a thyroidectomy? What are the consequences of this?

3.44 Describe the venous drainage of the thyroid gland.

4. Thorax

4.1   What movements do the ribs make during respiration?

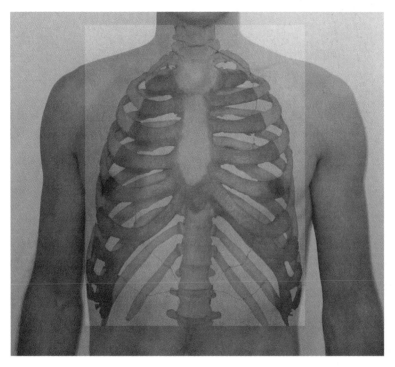

4.2   What are the surface markings of the heart? Indicate the cardiac borders on this chest.

4.3   What are the landmarks for inserting a chest drain?

4.4   Which layers would you go through when inserting a chest drain?

4.5   In which of these planes do the intercostal vessels run?

4.6   What is the blood supply to the oesophagus?

4.7   What constrictions occur in the oesophagus?

4.8   What are the surface markings of the tracheal carina?

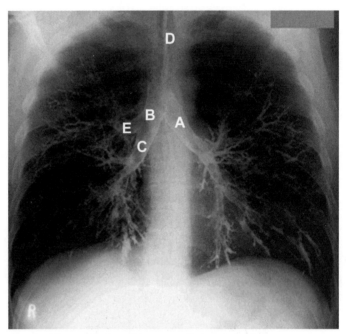

Reproduced with permission of Anatomedia Pty Ltd 2004
(www.anatomedia.com).

4.9

    a What investigation is this?
    b What structures are labelled A to E?

4.10 Which of the main bronchi (right or left) would an inhaled foreign object preferentially enter and why?

4.11 How many lobes do the lungs have?

4.12 How many bronchopulmonary segments does each lung have?

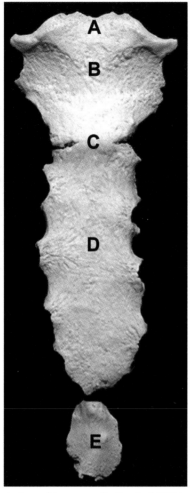

Reproduced with permission of Anatomedia Pty Ltd 2004
(www.anatomedia.com).

4.13 Name the different parts of the sternum labelled A to E.

4.14 Which thoracic vertebrae levels do the different parts of the sternum correspond to?

4.15 What is the motor nerve supply to the diaphragm? Which spinal levels does this nerve arise from?

4.16 Which nerves supply sensation to the diaphragm? Which spinal levels do these nerves arise from?

4.17 What is the clinical relevance of this sensory supply to the diaphragm?

4.18   What is the blood supply to the diaphragm?

4.19   What are the boundaries of the breast?

4.20   What is the blood supply to the breast?

4.21   What is the nerve supply to the breast and the nipple?

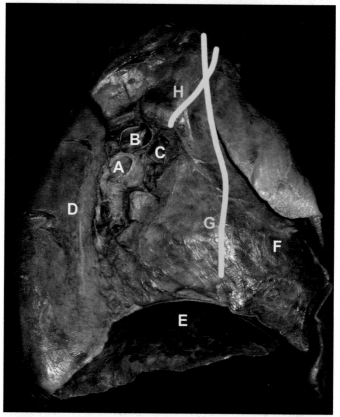

Reproduced with permission of Dartos Contractors (sca78548@bigpond.com.au).

4.22

    a   Which side is this lung from?
    b   Which structures cause the impressions on the lung labelled D to F?
    c   Which nerves run in the positions labelled G and H?
    d   What are the structures labelled A to C?

4.23   What are the surface markings of the lung fissures?

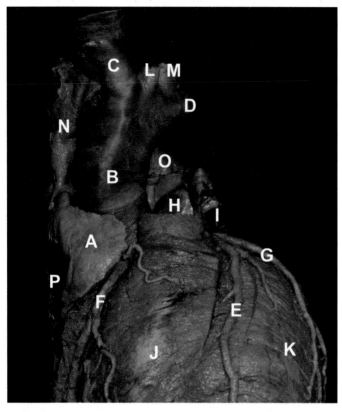

Reproduced with permission of Dartos Contractors (sca78548@bigpond.com.au).

4.24   Name the structures labelled A to P.

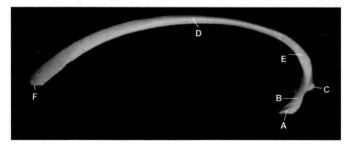

Reproduced with permission of Dartos Contractors (sca78548@bigpond.com.au).

4.25

    a Which bone is this?
    b Name points A to E.
    c What attaches to point F?
    d What articulates with point C?

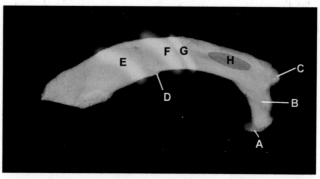

Reproduced with permission of Dartos Contractors (sca78548@bigpond.com.au).

4.26

   a Which bone is this?
   b Identify the bony points A to D.
   c What attaches to point D?
   d Which structures cause the grooves E and F?
   e Which structure runs in position G?
   f Which structures run above and below point B?
   g Which bone articulates with point A?
   h Which muscle attaches at point H?

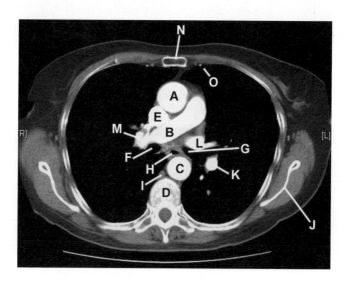

4.27

   a Which vertebral level is this CT scan?
   b Identify the structures labelled A to O.

4.28 Which structures would a needle pass through during subclavian vein cannulation?

4.29 Which structures are at risk during subclavian vein cannulation?

5. Abdomen and pelvis

5.1 Which abdominal and pelvic structures are intraperitoneal and which are retroperitoneal?

5.2 Which characteristics will help distinguish between jejunum and ileum?

5.3 Which characteristics will help distinguish between large bowel and small bowel?

5.4 Which structures pass through the diaphragm and at which levels?

5.5 What are the surface markings of these openings?

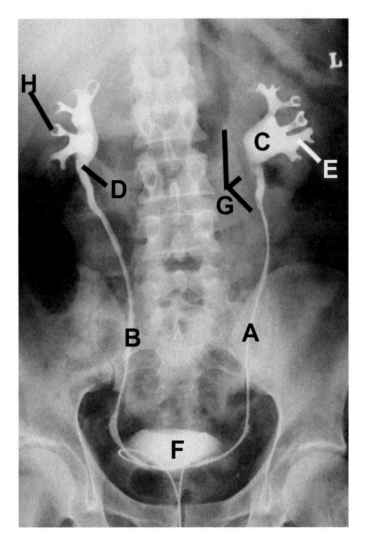

Reproduced with permission of Anatomedia Pty Ltd 2004
(www.anatomedia.com).

5.6

    a What is this investigation?
    b Identify the structures labelled A to F.
    c Which structure forms the faint line marked G?
    d What causes the cup-shaped impression labelled H?

5.7 Describe the course of the ureters. What are their relations?

5.8 What is the blood supply of the ureters?

5.9 What causes the constrictions of the ureter as seen on the intravenous urogram?

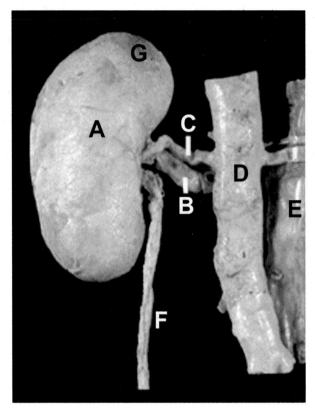

Reproduced with permission of Anatomedia Pty Ltd 2004 (www.anatomedia.com).

5.10

    a What organ is shown by A.
    b Identify the structures labelled B to F.
    c What organ is related to position G?
    d What is the usual order of structures at the hilum?
    e Therefore, which side is this organ from?
    f At what vertebral levels does this organ lie and what are its fascial layers?
    g What are the posterior relations of this organ?

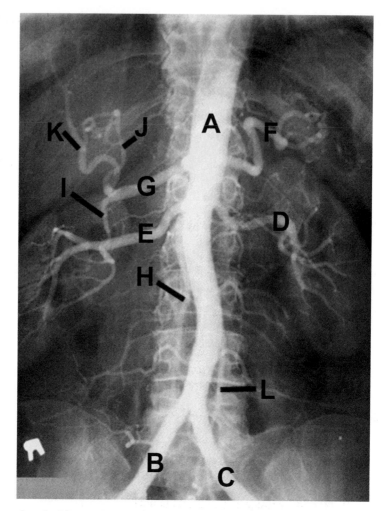

Reproduced with permission of Anatomedia Pty Ltd 2004
(www.anatomedia.com).

5.11

    a What is this investigation?
    b Identify the vessels labelled A to L.

5.12 Describe the course of the abdominal aorta.

5.13 Name the branches and describe the level at which they arise.

5.14 What is the transpyloric plane and what structures lie at this
level?

5.15 What layers would a surgeon cut through during a midline
laparotomy incision?

5.16 What structures make up the rectus sheath? Is this the same for the entire length?

5.17 What is the epiploic foramen (of Winslow)?

5.18 What are the boundaries of the epiploic foramen?

5.19 A finger can be inserted into the epiploic foramen and squeezed against a thumb anteriorly. What is this procedure called?

5.20 What structures are squeezed?

5.21 What is this procedure used for?

5.22 In the supine person, what part of the abdomen is the most dependent?

5.23 What clinical significance does this have?

5.24 What is the inguinal canal?

5.25 What are the boundaries of the inguinal canal?

5.26 What runs through the inguinal canal?

5.27 What is the mid-inguinal point?

5.28 What is the midpoint of the inguinal ligament?

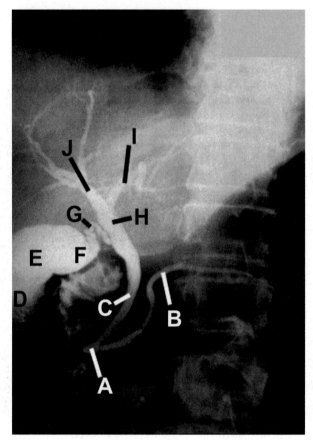

Reproduced with permission of Anatomedia Pty Ltd 2004
(www.anatomedia.com).

5.29

    a What is this investigation?
    b Name the parts labelled A to J.

5.30 Where does the common bile duct enter the bowel?

5.31 Which anatomical structure controls the secretion of bile?

5.32 What are the layers of the spermatic cord and scrotum and
which layers of the abdominal wall do they derive from?

5.33 What are the contents of the spermatic cord?

5.34 What is the fate of the right and left testicular veins?

5.35 What is the clinical relevance of this?

5.36   What is the lumbar plexus and which muscle is it related to?

5.37   Which nerves arise from the lumbar plexus?

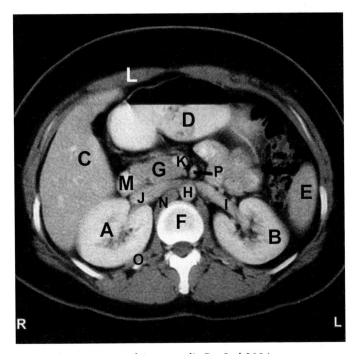

Reproduced with permission of Anatomedia Pty Ltd 2004
(www.anatomedia.com).

5.38

     a   What is this investigation and what level is it?
     b   Name the structures labelled A to P.

5.39   Describe the blood supply to the stomach.

5.40   Describe the venous drainage of the stomach.

5.41   Where are the sites of portosystemic anastomoses?

5.42   Describe the origin, path and branches of the superior
       mesenteric artery.

5.43   Describe the blood supply to the colon.

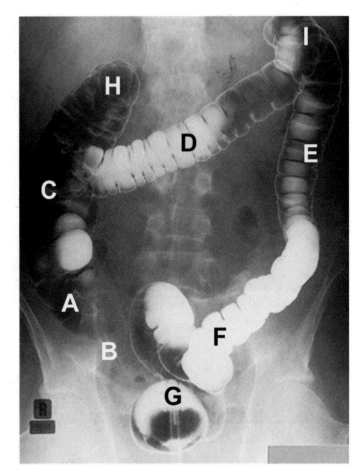

Reproduced with permission of Anatomedia Pty Ltd 2004
(www.anatomedia.com).

5.44   What is this investigation? Name the parts of the bowel labelled
A to I.

5.45   Which parts of the colon are retroperitoneal?

5.46   What is Calot's triangle? What forms the boundaries and what
are the contents?

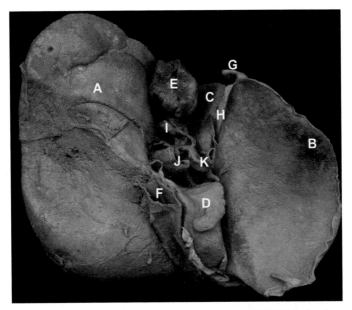

Reproduced with permission of Dartos Contractors (sca78548@bigpond.com.au).

5.47

a   What organ is this and from which angle is the photograph taken?

b   Name the different lobes labelled A to D.

c   What are the other structures labelled E to K?

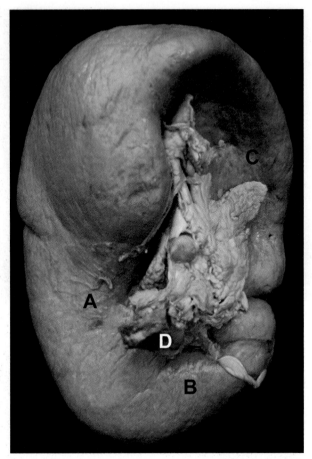

Reproduced with permission of Dartos Contractors (sca78548@bigpond.com.au).

5.48

    a What is this organ?

    b Which side of the body is it from?

    c What adjacent organs have left the impressions shown by A to D?

    d Is this organ normally palpable? If not, how much bigger must it be before it is palpable?

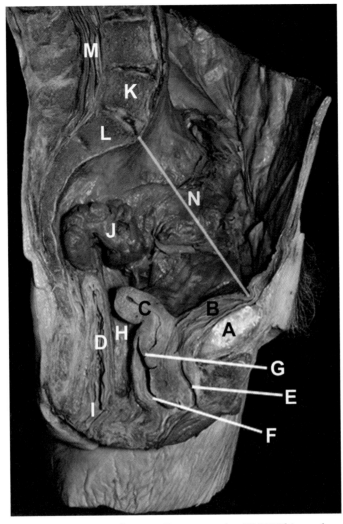

Reproduced with permission of Dartos Contractors (sca78548@bigpond.com.au).

5.49

    a Is this a male or female pelvis?
    b Name the structures labelled A to M.
    c Name the aperture labelled N.

5.50 Describe the different parts of the urethra in the male.

5.51 Which part of the urethra is responsible for control of the flow of urine?

5.52 Which is the most dilatable part of the urethra?

5.53 Which is the least dilatable part of the urethra?

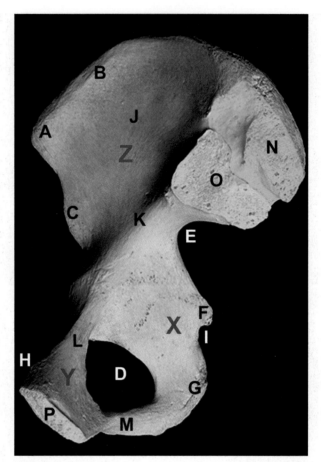

Reproduced with permission of Anatomedia Pty Ltd 2004
(www.anatomedia.com).

5.54

    a Which bone is this?
    b Identify points A to N?
    c Name the three bones labelled X to Z that form this bone?
    d Where do these three bones meet?
    e What attaches between points A and H?
    f What is represented by the blue shading at points O and P?
    g Which muscle originates from point J?

5.55   Describe the peritoneal relations of the rectum.

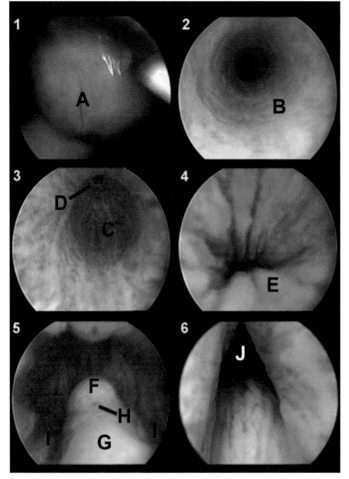

Reproduced with permission of Anatomedia Pty Ltd 2004
(www.anatomedia.com).

5.56

    a This is a series of photographs taken during a cystoscopy. Is it
       from a male or female?

    b Identify the structures labelled A to J.

    c What structure(s) opens into point H? What is its (their)
       function?

    d What structure(s) opens into point I?

6. Back

6.1   How many spinal nerves are there?

6.2   What landmarks are used when performing a lumbar puncture?

6.3   Which layers does the lumbar puncture needle pass through?

6.4   Describe the anatomy of the sympathetic chain.

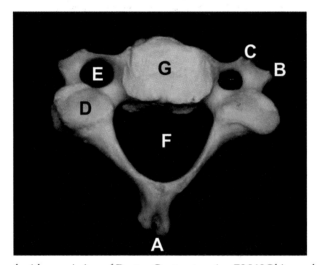

Reproduced with permission of Dartos Contractors (sca78548@bigpond.com.au).

6.5

    a Which region of the spine is this vertebra from?
    b Which characteristics help you identify its position?
    c Identify the parts labelled A to F.

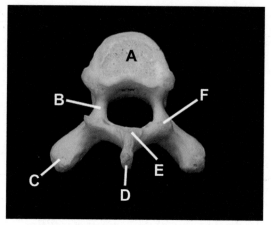

Reproduced with permission of Dartos Contractors (sca78548@bigpond.com.au).

6.6

a Which part of the spine is this vertebra from?
b What characteristics help you identify it?
c Identify points A to F.

Practice Viva Answers

1. Upper limb

1.1 Answer

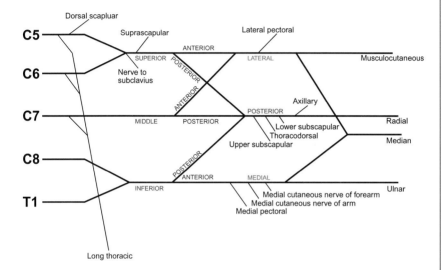

ROOTS   TRUNKS   DIVISIONS   CORDS   Branches

Practise drawing the basic pattern of the brachial plexus using a simplified line diagram like the one above. This should take 10 seconds. Then fill in the branches.

Three branches from the roots:
- Dorsal scapular nerve
- Long thoracic nerve
- Nerve to subclavius

One branch from the upper trunk:
- Suprascapular nerve

Three branches from the lateral cord:
- Lateral pectoral nerve
- Lateral root of median nerve
- Musculocutaneous nerve

Five branches from the posterior cord:
- Upper subscapular nerve
- Thoracodorsal nerve
- Lower subscapular nerve
- Axillary nerve
- Radial nerve

Five branches from the medial cord:
- Medial pectoral nerve
- Medial cutaneous nerve of arm
- Medial cutaneous nerve of forearm
- Medial root of median nerve
- Ulnar nerve

1.2 Answer

The posterior cord of the brachial plexus supplies the muscles that form the posterior border of the axilla, deltoid and posterior muscles of the arm and forearm.

The branches are the:
- Upper subscapular nerve – subscapularis
- Thoracodorsal nerve – latissimus dorsi
- Lower subscapular nerve – subscapularis and teres major
- Axillary nerve – deltoid, teres minor
- Radial nerve – triceps, all the extensor muscles of the forearm

1.3 Answer

The serratus anterior muscle is supplied by the long thoracic nerve (of Bell) (C567 Bells in heaven).

To test this muscle, ask the patient to face a wall and push both hands forward against it. Weakness of the serratus anterior muscle will cause characteristic winging of the scapula on that side.

1.4 Answer

None, as all the branches exit the brachial plexus before or after the divisions.

1.5 Answer

a This is a right humerus (with the humeral head pointing inwards the capitulum and trochlea pointing forwards).
b A Greater tubercle
   B Lesser tubercle
   C Intertubercular (bicipital) groove
   D Medial epicondyle
   E Lateral epicondyle
   F Trochlea
   G Capitulum
c H Pectoralis major
   I Latissimus dorsi
   J Teres major

Remember: Lady between two majors.

   K Deltoid

1.6 Answer

The forearm flexor muscles are arranged in two layers.

The *superficial layer* has five muscles arranged like five fingers radiating out from the common flexor origin of the medial epicondyle of the humerus (they all cross the elbow joint). From lateral to medial the muscles are:
- Pronator teres
- Flexor carpi radialis
- Flexor digitorum superficialis
- Palmaris longus
- Flexor carpi ulnaris

The *deep layer* has three muscles and they all originate in the forearm (none crosses the elbow joint):
- Flexor pollicis longus
- Flexor digitorum profundus
- Pronator quadratus

1.7 Answer

The forearm compartment flexor muscles are all supplied by the median nerve except for the flexor carpi ulnaris and the ulnar half of the flexor digitorum profundus muscles, which are supplied by the ulnar nerve.

1.8 Answer

The flexor digitorum profundus (FDP) tendon inserts into the base of the distal phalanx of the finger. It can be tested by asking the patient to flex the distal interphalangeal joint of that finger.

The flexor digitorum superficialis (FDS) tendon inserts into the base of the middle phalanx of the finger. Flexion at the proximal interphalangeal joint can be by the contraction of both the FDS and the FDP. Therefore, to test the function of the FDS muscle alone, the patient's FDP must be inactivated by holding the other fingers out straight and asking the patient to flex the unrestrained finger.

Remember: Superficialis splits in two to allow profundus passing through.

1.9 Answer

All of the intrinsic muscles of the hand are supplied by the ulnar nerve except for the lateral two lumbricals, opponens pollicis, abductor pollicis brevis and flexor pollicis brevis (LOAF muscles), which are supplied by the median nerve.

1.10 Answer

There are four lumbricals in each hand. They attach proximally to the FDP tendon, cross the radial side of the corresponding metacarpal phalangeal joint (MCPJ) and insert into the extensor expansion of that digit. They flex the MCPJ and extend the interphalangeal joints.

1.11 Answer

There are four dorsal interossei (between the bones). Each attaches proximally to adjacent sides of the two metacarpals it lies between (i.e. the fourth dorsal interossei attaches to the fourth and fifth metacarpals) and inserts distally onto the base of the proximal phalanx. The *D*orsal interossei *AB*duct the fingers. The little finger has a separate abductor (abductor digiti minimi).

There are three palmar interossei which attach proximally to the palmar aspect of the second, fourth and fifth metacarpals and insert distally into the base of the proximal phalanx of the corresponding digit. The *P*almar interossei *AD*uct the fingers.

Remember: PAD and DAB.

1.12 Answer

There are six compartments. From radial to ulnar:
1. Extensor pollicis brevis and abductor pollicis longus
2. Extensor carpi radialis longus and extensor carpi radialis brevis
3. Extensor pollicis longus
4. Extensor indicis proprius and extensor digitorum communis
5. Extensor digiti minimi
6. Extensor carpi ulnaris

1.13 Answer

a This area is known as the anatomical snuff box.
b The anatomical snuff box is bounded on the radial side by the abductor pollicis longus (A) and extensor pollicis brevis (B) tendons. On the ulnar side is the extensor pollicis longus (C) tendon. The floor is the scaphoid and trapezium bones.
c The contents of the anatomical snuff box are the radial artery, radial nerve and extensor carpi radialis longus and brevis tendons.

1.14 Answer

The rotator cuff is made up of four muscles:
- Supraspinatus
- Infraspinatus
- Teres minor
- Subscapularis

Remember: SITS.

These four muscles blend their tendinous insertions into the articular capsule of the glenohumeral joint. This arrangement holds the humerus tightly in position, giving greater stability to the shoulder joint, but still allowing a large range of movement.

1.15 Answer

The axilla is a pyramidal, intermuscular space with the following boundaries:
- Apex: cervicoaxillary canal (the convergence of clavicle, scapula and first rib)
- Anterior: pectoralis major and minor muscles
- Base: axillary fascia
- Posterior: subscapularis, teres major, latissimus dorsi muscles (superior to inferior)
- Medial: upper three ribs, intercostal spaces and serratus anterior muscle
- Lateral: intertubercular groove of humerus, short head of biceps, coracobrachialis tendon

1.16 Answer

The contents of the axilla are:
- Axillary artery and its branches
- Axillary vein and its tributaries
- Axillary lymph nodes:
    Level 1 (below pectoralis minor)
    Level 2 (behind pectoralis minor)
    Level 3 (above pectoralis minor)
- Brachial plexus – cords and branches
- Fat

1.17 Answer

The axillary artery is divided into three parts:
- First part is medial to the pectoralis minor and has one branch:
    Superior thoracic artery
- Second part is behind the pectoralis minor and has two branches:
    Thoracoacromial trunk
    Lateral thoracic artery
- Third part is lateral to the pectoralis minor and has three branches:
    Subscapular artery
    Anterior circumflex humeral artery
    Posterior circumflex humeral artery

1.18 Answer

The clavipectoral fascia is a thin layer of fibrous tissue that surrounds the pectoralis minor muscle. It attaches superiorly to the clavicle and inferiorly to the axillary fascia. It is thus the suspensory ligament of the axilla below the pectoralis minor.

Four structures pierce the clavipectoral fascia:
- Two structures passing in:
    Cephalic vein
    Lymphatic vessels
- Two structures passing out:
    Lateral pectoral nerve
    Thoracoacromial trunk

1.19 Answer

The quadrangular space is formed:
- Laterally by the humerus
- Medially by the long head of triceps
- Superiorly by the teres minor
- Inferiorly by the teres major

[handwritten: T Mm / LB   H / Tmaj.]

Passing through this space is the axillary nerve and the posterior circumflex humeral artery.

1.20 Answer

The nerves that can be damaged are the:
- Axillary nerve as it passes close to the neck of the humerus in the quadrangular space
- Radial nerve as it winds around the shaft of the humerus at the junction between the proximal two-thirds and distal one-third
- Ulnar nerve as it passes behind the medial epicondyle
- Median nerve in a supracondylar fracture

1.21 Answer

a   A   Biceps brachii
     B   Brachioradialis
     C   Pronator teres
b   D   Cephalic vein
     E   Basilic vein
     F   Median cubital vein
     G   Radial artery
     H   Brachial artery
     I   Cephalic vein
     J   Lateral cutaneous nerve of forearm (termination of musculocutaneous nerve)
     K   Medial cutaneous nerve of forearm (branch of medial cord of brachial plexus)
     L   Biceps tendon

1.22 Answer

The cubital fossa is a triangular intermuscular space bounded:
- Superiorly: the interepitrochlear line
- Medially: the pronator teres
- Laterally: the brachioradialis
- Floor: the brachialis
- Roof: the deep fascia of the forearm

1.23 Answer

From lateral to medial the contents of the cubital fossa are the:
- Biceps *T*endon
- Brachial *A*rtery (with venae comitantes)
- Median *N*erve

Remember: TAN.

1.24 Answer

a *Proximal row*:
   A Scaphoid
   B Lunate
   C Triquetrum
   D Pisiform
   Distal row:
   E Hamate
   F Capitate
   G Trapezoid
   H Trapezium
b X marks the hook of the hamate.
c Y marks the proximal interphalangeal joint of the index finger.

1.25 Answer

The flexor retinaculum attaches to the:
- Hook of the hamate and lateral ridge of the trapezium distally
- Pisiform and the tubercle of the scaphoid proximally

The contents of the carpal tunnel are:
- Four flexor digitorum superficialis tendons
- Four flexor digitorum profundus tendons
- Median nerve
- Flexor pollicis longus tendon
- Flexor carpi radialis tendon

1.26 Answer

During open carpal tunnel decompression the following layers are cut:
- Skin
- Subcutaneous fat
- Palmar fascia
- (Palmaris brevis muscle, occasionally)
- Flexor retinaculum

It is important to divide the entire length of the flexor retinaculum.

At risk of damage in open carpal tunnel release are the:
- Palmar cutaneous branch of the median nerve (sensation to the thenar eminence)
- Recurrent branch of the median nerve (motor branch to the thenar muscles)
- Ulnar nerve as it passes through the flexor retinaculum (with an incision too ulnar)
- Median nerve
- Superficial palmar arch
- Flexor tendons passing through the carpal tunnel

1.27 Answer

This is the C6 dermatome.

1.28 Answer

a A Flexor carpi radialis
  B Palmaris longus
  C Flexor carpi ulnaris
  D Flexor pollicis longus
  E Brachioradialis
  F Flexor digitorum superficialis
  G Abductor pollicis brevis
  H Flexor pollicis brevis
  I Flexor digiti minimi
  J Abductor digiti minimi
b K Flexor pollicis longus
  L Flexor digitorum profundus to little finger
  M Flexor digitorum superficialis to ring finger
c At point X the flexor digitorum superficialis (FDS) tendon splits to insert into the middle phalanges. The flexor digitorum profundus tendon passes through this split in the FDS to insert into the base of the distal phalanges.

1.29 Answer

To prevent the flexor tendons from bow stringing there are a series of fascial coverings that anchor the tendon to the bony skeleton whilst still allowing the tendons to glide smoothly.

At the wrist there is the flexor retinaculum and on the fingers the pulleys. There are the annular (A1–5) and the cruciate pulleys (C1–3):
- A1 pulley is over the metacarpal phalangeal joint
- A3 pulley is over the proximal interphalangeal joint
- A5 pulley is over the distal interphalangeal joint
- A2 pulley is over the proximal phalanx (most important pulley)
- A4 pulley is over the middle phalanx (second most important pulley)
- C1 pully lies between A2 and A3
- C2 pulley lies between A3 and A4
- C3 pully lies between A4 and A5

From proximal to distal the pulleys are: A1, A2, C1, A3, C2, A4, C3, A5.

1.30 Answer

The musculocutaneous nerve is the continuation of the lateral cord of the brachial plexus.

1.31 Answer

The musculocutaneous nerve supplies the:
- Biceps brachii
- Brachialis
- Corocobrachialis

Remember: The musculocutaneous nerve is the BBC nerve.

1.32 Answer

Yes, it does. Hence the name musculo*cutaneous*. After giving off its motor branches it crosses the lateral border of the biceps tendon and continues as the lateral cutaneous nerve of the forearm. This supplies sensation to the radial half of the volar forearm and also some of the dorsal surface.

2. Lower limb

2.1 Answer

There are four compartments in the leg:
- Anterior
- Lateral
- Superficial posterior
- Deep posterior

2.2 Answer

a A Tibia
   B Fibula
   C Soleus
   D Gastrocnaemius
   E Plantaris tendon
   F Interosseus membrane
   G Flexor hallucis longus
   H Tibialis posterior
   I Flexor digitorum longus
   J Tibialis anterior
   K Peroneus longus
b L Anterior tibial artery
   M Peroneal artery
   N Posterior tibial artery
c O Long saphenous vein
   P Short saphenous vein
d The saphenous nerve runs with the long saphenous vein (O) whilst the sural nerve runs with the short saphenous vein (P).
e The deep peroneal nerve runs in the anterior compartment with the anterior tibial artery (L). The tibial nerve runs in the deep posterior compartment with the posterior tibial artery (N).

2.3 Answer

The following muscles lie in the anterior compartment:
- Tibialis anterior
- Extensor digitiorum longus
- Extensor hallucis longus
- Peroneus tertius

2.4 Answer

The patient will be left with a foot drop.

2.5   Answer

- Medially: the medial border of the adductor longus
- Laterally: the medial border of the sartorius
- Superiorly: the inguinal ligament
- Roof: the fascia lata
- Floor: iliacus, psoas tendon, pectineus and adductor longus

2.6   Answer

From lateral to medial: femoral nerve, femoral artery, femoral vein and lymph nodes.

Remember: NAVY: Nerve, Artery, Vein, Y fronts.

2.7   Answer

The femoral sheath is a continuation of the extraperitoneal fascia – formed anteriorly by the transversalis fascia and posteriorly by the ileopsoas fascia. It is a fascial tube that extends for 3 cm below the inguinal ligament and surrounds the femoral artery, femoral vein and femoral canal containing the lymphatic vessels. It does not contain the femoral nerve.

2.8   Answer

The femoral canal is the most medial of the compartments of the femoral sheath. It contains fat, lymphatic vessels and Cloquet's node. Superiorly the femoral canal opens into the femoral ring.

2.9   Answer

The femoral ring is the superior mouth of the femoral canal. Its boundaries are:
- Anterior: inguinal ligament
- Posterior: superior ramus of the pubis covered in the pectineal ligament (of Astley Cooper)
- Medially: lacunar ligament
- Lateral: fascia around the femoral vein
  It is where a femoral hernia would occur.

2.10   Answer

The dorsalis pedis pulse is palpated on the dorsum of the foot just lateral to the extensor hallucis longus tendon over the cuneiform bones.

   The posterior tibial pulse is palpable halfway between the posterior border of the medial malleolus and the tendo-Achilles.

2.11   Answer

a   This is a digital subtraction angiogram (DSA) of the left leg.
b   A   Common femoral artery
    B   Superficial femoral artery
    C   Profunda femoris artery
    D   Popliteal artery

E   Anterior tibial artery (*Note* the perpendicular course as it pierces the interosseous membrane)
F   Tibioperoneal trunk
G   Peroneal artery
H   Posterior tibial artery
I   Dorsalis pedis artery
J   Lateral plantar artery

2.12   Answer

The external iliac artery changes its name to the femoral artery after passing under the inguinal ligament. The femoral (or common femoral) artery gives off a deep branch called the profunda femoris that supplies all the thigh muscles. The continuation of the femoral artery is known by vascular surgeons as the superficial femoral artery. This vessel then becomes the popliteal artery after passing through the adductor hiatus of the adductor magnus tendon. In the popliteal fossa, it gives off five genicular branches (superior medial, superior lateral, inferior medial, inferior lateral and middle). The popliteal artery ends by dividing into the anterior tibial artery (smaller) and tibioperoneal trunk (larger). The anterior tibial artery pierces the interosseous membrane and then runs in the anterior compartment of the leg, supplying all the muscles of this compartment. This artery becomes the dorsalis pedis artery after crossing the ankle joint.

The tibioperoneal trunk divides into the peroneal artery (larger branch) that runs close to the fibula and the posterior tibial artery that runs in the posterior compartment close to the tibial nerve. At the ankle, the posterior tibial artery runs behind the medial malleolus and then divides into the medial and lateral plantar arteries. The lateral plantar artery anastomoses with the dorsalis pedis artery at the plantar arch.

2.13   Answer

The long saphenous vein starts in the foot at the confluence of the dorsal venous arch and the dorsal vein of the great toe. It passes just anterior to the medial malleolus to climb up the calf, just posterior to the palpable border of the tibia. It passes a hand's breadth behind the medial border of the patella, continues up the medial aspect of the thigh and then pierces the fascia lata to drain into the femoral vein at the saphenofemoral junction (approximately 2 cm below and lateral to the pubic tubercle).

There are numerous perforating veins connecting the deep and superficial venous systems. As a rough guide, these perforators are: 1 cm below, 1 cm above and 10 cm above the medial malleolus, one in the middle of the calf and one just below the knee. In reality, the position of the perforators is quite variable (hence the need for Doppler marking prior to varicose vein surgery).

2.14 Answer

The saphenous nerve runs close to the long saphenous vein below the knee. It may be damaged during stripping of the inferior part of the long saphenous vein when treating varicose veins.

2.15 Answer

The saphenous nerve is a branch of the femoral nerve and passes through the subsartorial canal with the femoral vessels.

2.16 Answer

The saphenous nerve supplies sensation to the medial aspect of the calf and foot.

2.17 Answer

The subsartorial (adductor or Hunter's) canal is an intermuscular tunnel beneath the middle third of the sartorius muscle. It ends at the adductor hiatus, an opening in the adductor magnus tendon.

2.18 Answer

- Lateral: vastus medialis
- Posterior: adductor longus and adductor magnus
- Anterior: sartorius

2.19 Answer

The contents of the subsartorial canal are the femoral vessels as they pass from the anterior thigh to become the popliteal vessels:
- Superficial femoral artery (anterior)
- Femoral vein (posterior)
- Saphenous nerve (*Note*: This nerve does not exit through the adductor hiatus but perforates between the gracilis and sartorius to run with the long saphenous vein)
- Nerve to vastus medialis (*Note*: Again, this nerve does not exit through the adductor hiatus)

2.20 Answer

a This is the left knee joint.

Note: the medial condyle or plateau of the tibia is larger than the lateral side. The lateral femoral condyle has a greater prominence on the side of the patella groove to prevent the patella being dislocated laterally with every contraction of the quadriceps – this muscle has a lateral pull.

b A Lateral condyle of femur
B Medial condyle of femur
C Intercondylar fossa
D Patella groove

c From anterior to posterior the following structures attach into the tibial plateau:
E Anterior horn of medial meniscus
F Anterior cruciate ligament
G Anterior horn of lateral meniscus
H Posterior horn of lateral meniscus
I Posterior horn of medial meniscus
J Posterior cruciate ligament

2.21 Answer

The cruciate ligaments attach the tibia to the femur and are named according to their position of origin on the tibia.

The *anterior cruciate ligament* attaches to the anterior part of the intercondylar area of the tibia, just behind the medial meniscus. It runs superiorly and posteriorly to attach into the lateral condyle/intercondylar notch of the femur. It prevents the tibia from being displaced anteriorly from the femur and prevents hyperextension of the knee.

The anterior cruciate ligament is tested by assessing the amount of anterior movement of the tibia relative to the femur with the knee flexed to 90° (anterior drawer test).

The *posterior cruciate ligament* attaches to the posterior aspect of the intercondylar area of the tibia. It runs superiorly and anteriorly to attach into the medial condyle/intercondylar notch of the femur. It prevents the tibia being displaced posteriorly from the femur and prevents hyperflexion of the knee.

The posterior cruciate ligament is tested by assessing the amount of posterior movement of the tibia relative to the femur with the knee flexed to 90° (posterior drawer test).

2.22 Answer

The posterior cruciate ligament is stronger than the anterior.

2.23 Answer

a This is the medial aspect of the left tibia.
b This is the pes anserinus (from the Latin for goose's foot).
c From anterior to posterior:
A Sartorius
B Gracilis
C Semitendinosus

Remember: Say Grace before Tea.

d The tibial (or medial) collateral ligament.

2.24 Answer

a This is the medial malleolus.
b Passing posterior to the medial malleolus, under the flexor retinaculum are (from anterior to posterior):
• Tibialis posterior tendon

- Flexor Digitorum longus tendon
- Posterior tibial Artery
- Venae comitantes of the posterior tibial artery
- Tibial Nerve
- Flexor Hallucis longus tendon

Remember: Tom, Dick And Very Naughty Harry.

c This is the tendo calcaneus or Achilles tendon.
d The three muscles of the superficial posterior compartment of the leg insert into the calcaneus via the Achilles tendon. They are the:
  - Gastrocnaemius (medial and lateral heads)
  - Soleus
  - Plantaris
e All the superficial posterior compartment muscles are supplied by the tibial nerve.
f The bones that can be palpated along the medial border of the foot are:
  C Distal phalanx of great toe
  D Proximal phalanx of great toe
  E First metatarsal
  F Medial cuneiform
  G Navicular
  H Talus
  I Calcaneus

2.25 Answer

a This is the posterior aspect of a right femur.
b A Head of femur
  B Neck of femur
  C Greater trochanter
  D Lesser trochanter
  E Intertrochanteric crest
  F Adductor tubercle
  G Medial condyle
  H Lateral condyle
  I Intercondylar fossa
  J Linea aspera
  K Gluteal tuberosity

2.26 Answer

The femoral head receives a small amount of blood from the artery within the ligamentum teres, but this is usually inadequate alone.

The majority of the blood supply comes from the extracapsular arterial ring that lies around the base of the femoral neck. This anastomosis receives branches from the medial and lateral circumflex femoral arteries and a smaller contribution from the superior and inferior gluteal arteries. Small vessels from this anastomosis pass through the medulla of the femoral neck to supply the femoral head.

2.27 Answer

Intracapsular fractures of the neck of the femur disrupt the intraosseous blood supply to the femoral head. The blood supply through the ligamentum teres is usually inadequate, thus leading to avascular necrosis of the head of the femur.

2.28 Answer

The anterior thigh muscles are the hip flexors and the knee extensors. They are the:
- Pectinius
- Iliopsoas
- Tensor fascia lata
- Sartorius
- Quadriceps femoris
    Rectus femoris
    Vastus lateralis
    Vastus intermedialis
    Vastus medialis

2.29 Answer

The medial (or adductor) compartment of the thigh is supplied by the obturator nerve.

2.30 Answer

a A Biceps femoris
    B Lateral head of gastrocnaemius
    C Medial head of gastrocnaemius
    D Semimembranosis
    E Semitendonosis
    F Common peroneal nerve
    G Tibial nerve
    H Sural nerve
b The common peroneal nerve (F) divides into the superficial and deep peroneal nerves.

2.31 Answer

The popliteal fossa is a diamond-shaped intermuscular region on the posterior aspect of the knee. The boundaries are:
- Superomedially: semimembranosis and semitendinosis muscles
- Inferomedially: medial head of gastrocnaemius muscle
- Superolaterally: biceps femoris muscle
- Inferolaterally: lateral head of gastrocnaemius and plantaris muscles
- Floor: the capsule of the knee joint, popliteus muscle and femur
- Roof: popliteal fascia

2.32 Answer

The contents of the popliteal fossa are (from deep to superficial):
- Popliteal artery
- Popliteal vein
- Tibial nerve
- Common peroneal nerve

The popliteal fossa also contains:
- Fat
- Lymph nodes
- Termination of the short saphenous vein as it enters the popliteal vein
- Sural nerve
- Popliteus bursa
- The five genicular branches of the popliteal artery

2.33 Answer

a   This is the right gluteal region.

Note: Whenever shown a photograph or specimen of the gluteal region use the piriformis muscle as a landmark to orientate yourself.

b   A   Piriformis
   B   Superior gemellus
   C   Obturator internus
   D   Inferior gemellus
   E   Gluteus medius
   F   Gluteus maximus
   G   Quadratus femoris
c   The superior gluteal nerve exits the pelvis above the piriformis (A).
d   The superior gluteal nerve supplies the gluteus medius and gluteus minimus.
e   They are abductors of the hip joint.
f   The patient would have a Trendelenburg gait.
g   H   Sciatic nerve
   I   Pudendal nerve
   J   Superior gluteal nerve
   K   Inferior gluteal nerve
h   L   Inferior gluteal artery
   M   Superior gluteal artery
i   The superior and inferior gluteal arteries are branches of the internal iliac artery.
j   The superior and inferior gluteal arteries exit the pelvis through the greater sciatic foramen.
k   N marks the greater trochanter of the femur. This is the insertion of the gluteus medius and minimus, piriformis, superior and inferior gemelli, and obturator internus.

2.34 Answer

The following structures are cut through during the posterior approach to the hip joint:
- Skin
- Subcutaneous fat
- Gluteal fascia
- Gluteus maximus
- Short external rotator muscles
- Hip joint capsule

2.35 Answer

The nerves at risk are the sciatic nerve and superior gluteal nerve.

2.36 Answer

Deep to the plantar fascia there are four muscle layers. From superficial to deep they are the:
- Abductor hallucis, abductor digiti minimi and flexor digitorum brevis
- Quadratus plantae and four lumbricals (and tendons of the flexor digitorum longus and flexor hallucis longus)
- Flexor hallucis brevis, adductor hallucis and flexor digiti minimi
- Three plantar interossei and four dorsal interossei (and the tendons of the tibialis posterior and peroneus longus)

The medial and lateral plantar nerves and medial and lateral planter arteries run between the first and second layers of muscles. The deep plantar arch is between the third and fourth layers.

2.37 Answer

The lateral cutaneous nerve of the thigh supplies sensation to the outside of the thigh.

2.38 Answer

This is the L2 and L3 dermatome.

2.39 Answer

1–2 cm medial and inferior to the anterior superior iliac spine as the nerve passes below the inguinal ligament.

2.40 Answer

The sensory nerve supply is the deep peroneal nerve.

2.41 Answer

The nerve roots of the sciatic nerve are L4,5, S1,2,3.

2.42 Answer

The terminal branches of the sciatic nerve are the tibial nerve and the common peroneal nerve.

2.43 Answer

- L5 sensation: dorsum of the foot
- L5 motor: test extensor hallucis longus (lift up big toe)

3. Head and neck

3.1 Answer

The deep fascia of the neck has four layers:
- The investing layer of deep cervical fascia – this is the most superficial layer and surrounds the entire neck. It is deep to the platysma muscle. It splits to enclose the sternocleidomastoid muscles anteriorly and the trapezius muscles posteriorly.
- The pre-vertebral layer of deep cervical fascia lies posteriorly and surrounds the vertebral column and the associated muscles (scalene muscles, longus colli, etc).
- The pre-tracheal layer of deep cervical fascia lies anteriorly and surrounds the trachea, oesophagus and thyroid gland.
- The carotid sheath is another anterior (but paired) sheath of deep cervical fascia that lies on either side of the pre-tracheal fascia. It surrounds the carotid artery medially, internal jugular vein laterally and vagus nerve in between. It also contains the ansa cervicalis and some lymph nodes.

3.2 Answer

The posterior triangle of the neck is bordered by:
- Anteriorly: posterior border of the sternocleidomastoid muscle
- Posteriorly: anterior border of the trapezius muscle
- Inferiorly: middle third of the clavicle
- Roof: investing layer of deep cervical fascia
- Floor: pre-vertebral fascia over the top of the following muscles:
    Splenius capitis
    Levator scapulae
    Scalenus anterior
    Scalenus medius
    Scalenus posterior

3.3 Answer

The *sternocleidomastoid* turns the head to the side opposite the muscle. To test the left sternocleidomastoid ask the patient to turn his/her head to the right.

The *trapezius* elevates the scapula. To test the muscle, ask the patient to shrug his/her shoulders.

3.4 Answer

The contents of the posterior triangle of the neck are:

Nerves
- Spinal accessory nerve
- Cervical plexus (lesser occipital, greater auricular, transverse cervical, supraclavicular)
- Brachial plexus (superior, middle and inferior trunks)

Arteries
- Third part of the subclavian artery

- Transverse cervical artery) ⎫ both branches of the thyrocervical
- Suprascapular artery) ⎬ trunk, a branch of the first part of
                 ⎭ the subclavian artery

- Occipital artery – at the apex of the posterior triangle

Veins
- External jugular vein

Lymph nodes

3.5 Answer

Damage to the spinal accessory nerve would result in paralysis of the ipsilateral trapezius and sternocleidomastoid muscles.

3.6 Answer

During tracheostomy, the following layers are traversed:
- Skin
- Subcutaneous fat
- Platysma
- Investing layer of deep cervical fascia
- Strap muscles (sternohyoid and sternothyroid) – these are usually pulled aside rather than cut
- Pre-tracheal fascia
- Thyroid isthmus (ligated and divided)
- Trachea

3.7 Answer

a A Supraorbital foramen – supraorbidal nerve, artery and vein
  B Infraorbital foramen – infraorbital nerve, artery and vein
  C Mental foramen – mental nerve, artery and vein
b U Sphenoid
  V Mandible
  W Zygoma
  X Frontal bone
  Y Nasal bone
  Z Maxilla

3.8   Answer

The sphenoid articulates with eight other bones:
- Temporal
- Parietal
- Frontal
- Vomer
- Occipital
- Zygomatic
- Palatine
- Ethmoidal

3.9   Answer

a

A	Foramen ovale	Mandibular division of the trigeminal nerve
B	Carotid canal	Internal carotid artery Sympathetic plexus
C	Jugular foramen	Internal jugular vein (formed by the inferior petrosal and sigmoid sinuses) Glossopharyngeal nerve Vagus nerve Accessory nerve
D	Stylomastoid foramen	Facial nerve Stylomastoid artery
E	Foramen magnum	Medulla and surrounding meninges Spinal roots of accessory nerve Anterior and posterior spinal and vertebral arteries
F	Foramen spinosum	Middle meningeal vessels
G	Foramen lacerum	Internal carotid artery (passes into foramen lacerum from carotid canal)

b
- H Occipital condyle
- I External occipital protuberance
- J Styloid process
- K Vomer

3.10   Answer

The facial nerve (or seventh cranial nerve) emerges from the junction of the pons and medulla. It traverses the posterior cranial fossa, runs within its own canal in the temporal bone and then exits the skull at the stylomastoid foramen.

Within the facial canal the facial nerve gives off the:
- Greater petrosal nerve. These are parasympathetic fibres that then hitch a ride with the maxillary division of the trigeminal nerve – CN V2 – to supply the lacrimal gland
- Nerve to stapedius
- Chorda tympani. These are special sensory and parasympathetic fibres that hitch a lift with the lingual nerve – a branch of the mandibular division of the trigeminal nerve (CN V3) to provide

taste to the anterior two-thirds of the tongue and innervation to the submandibular and sublingual salivary glands
- Auricular branch supplies sensation to a small area of skin around the external auditory meatus

After leaving the skull the facial nerve gives off the posterior auricular nerve (which supplies the occipital belly of occipitofrontalis) and a branch that supplies the stylohyoid muscle and the posterior belly of digastric then enters the parotid gland. Within the gland, the nerve divides into five branches: temporal, zygomatic, buccal, mandibular and cervical. These branches supply the muscles of facial expression.

Remember: The function of the facial nerve from the mnemonic: face, ear, taste, tear.

3.11 Answer

- Temporal: raise your eyebrows
- Zygomatic: close your eyes tightly
- Buccal: blow your cheeks out
- Mandibular: show me your bottom teeth
- Cervical: tense the skin on your neck/chin (platysma)

3.12 Answer

Four nerves contribute sensation to the ear:
- Great auricular nerve (branch of the cervical plexus)
- Auriculotemporal branch of the mandibular nerve
- Auricular branch of the vagus nerve
- Branches of the facial nerve (branches from the tympanic plexus)

3.13 Answer

A Helix
B Antihelix
C Superior crus
D Inferior crura
E Triangular fossa
F Scapha
G Concha
H Tragus
I Intertragic notch
J Antitragus
K Lobule
L External acoustic meatus

3.14 Answer

The network of cerebral veins drain into the venous sinuses. These are channels between the dura and the periosteum that contain no valves.
- Superior sagittal sinus – this lies in the superior falx cerebri and runs from the crista gali anteriorly to the internal occipital protuberance posteriorly (confluence of sinuses)

- Inferior sagittal sinus – this lies in the inferior border of the falx cerebri and joins the great cerebral vein to become the straight sinus
- Straight sinus – this is formed by the confluence of the inferior sagittal sinus and the great cerebral vein and runs inferioposteriorly to the internal occipital protuberance (confluence of sinuses)
- Transverse sinuses – these (right and left) pass laterally from the internal occipital protuberance (confluence of sinuses) to become the sigmoid sinuses
- Sigmoid sinuses – these curve medially and then exit through the jugular foramina to become the internal jugular veins
- Occipital sinus – this runs superiorly from the epidural plexus of veins to the confluence of sinuses
- Cavernous sinuses – these lie on either side of the sella turcica. These sinuses drain the ophthalmic, sphenoparietal and middle cerebral veins into the superior and inferior petrosal sinuses
- Superior petrosal sinuses – these run from the cavernous sinuses to the junction of the transverse and sigmoid sinuses
- Inferior petrosal sinuses – these run from the cavernous sinuses to empty directly into the internal jugular veins
- Basilar sinuses – these connect the inferior petrosal sinuses with the epidural plexus of veins

The emissary veins connect the intra- and extra-cranial veins.

3.15 Answer

The cavernous sinuses are paired venous sinuses, about 2 cm long on either side of the sella turcica and lateral to the sphenoid air sinuses, immediately posterior to the optic chiasm. Draining into these sinuses are the superior and inferior ophthalmic veins, the superficial middle cerebral vein and the sphenoparietal sinuses. The cavernous sinuses communicate with each other via the intercavernous sinuses and drain into the inferior and superior petrosal sinuses.

The cavernous sinus contains the internal carotid artery, occulomotor nerve, troclear nerve, ophthalmic and maxillary divisions of the trigeminal nerve, abducens nerve and sympathetic plexus.

Infections or tumours of the face may spread through the facial veins and ophthalmic veins (valveless) into the cavernous sinuses causing thrombosis. This condition manifests as a swollen, painful, venous congested ipsilateral eye, progressive loss of vision and the development of third, fourth, fifth and sixth cranial nerve palsies. The condition may quickly spread to the contralateral sinus.

3.16 Answer

  a A Internal carotid artery
      B Anterior communicating artery
      C Anterior cerebral artery
      D Middle cerebral artery
      E Basilar artery
  b This is the circle of Willis.

3.17 Answer

C1 = Hard palate
C2 = Angle of mandible
C3 = Hyoid bone
C4 = Superior thyroid notch
C5 = Thyroid cartilage
C6 = Cricoid cartilage
C7 = Upper tracheal rings

3.18 Answer

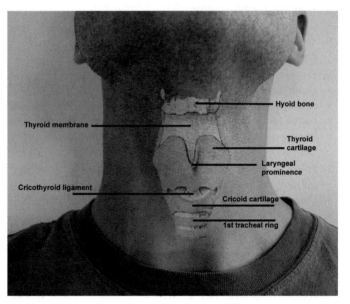

Hyoid bone

Thyroid membrane

Thyroid cartilage

Laryngeal prominence

Cricothyroid ligament

Cricoid cartilage

1st tracheal ring

3.19 Answer

From inferior to superior the branches of the external carotid artery are:

- Ascending pharyngeal
- Superior thyroid
- Lingual
- Facial
- Occipital
- Posterior auricular
- Superficial temporal
- Maxillary

3.20 Answer

The internal carotid artery has no branches in the neck but after passing through the cavernous sinus divides into the anterior and middle cerebral artery and also gives off the smaller posterior communicating artery.

3.21 Answer

The common carotid artery bifurcates into the internal and external carotid arteries at the level of C4.

3.22 Answer

The maxillary artery is divided into three parts by its relation to the lateral pterygoid muscle.

First part (inferior to lateral pterygoid) has five branches:
- Deep auricular artery
- Anterior tympanic artery
- Middle meningeal artery
- Accessory meningeal artery
- Inferior alveolar artery

Second part (behind lateral pterygoid) has four branches:
- Deep temporal artery
- Pterygoid artery
- Masseteric artery
- Buccal artery

Third part (superior to lateral pterygoid) has six branches:
- Posterior superior alveolar artery
- Infraorbital artery
- Descending palatine artery
- Artery of pterygoid canal
- Pharyngeal artery
- Sphenopalatine artery

3.23 Answer

There are 12 cranial nerves:
- Olfactory
- Optic
- Oculomotor
- Trochlea
- Trigeminal
- Abducens
- Facial
- Vestibulocochlear
- Glossopharyngeal
- Vagus
- Accessory
- Hypoglossal

3.24 Answer

The fifth cranial or trigeminal nerve supplies sensation to the face and motor supply to the muscles of mastication. There are three divisions:
- Ophthalmic
- Maxillary
- Mandibular

3.25 Answer

The *ophthalmic division* supplies sensation to the forehead, upper eyelids, eye, anterior nose and nasal mucosa.

The *maxillary division* supplies sensation to the cheek, lower eyelid, lateral nose, upper teeth, upper lip and maxillary sinuses.

The *mandibular division* supplies sensation to the skin over the mandible, lower teeth and lip, temporal skin and lower oral cavity.

3.26 Answer

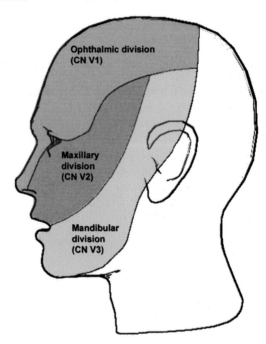

3.27 Answer

a A Coronal suture
  B Squamous suture
  C Lamboid suture
  D Zygomaticofacial suture
  E Sagittal suture
b This is the pterion. It is the junction of the frontal, parietal, temporal and sphenoid bones.
c The frontal branch of the middle meningeal artery runs behind it.
d The pterion is a weak point of the skull and may be fractured by a blow to the temporal region, e.g. from a cricket ball. A fracture may lacerate the middle meningeal artery leading to an extradural haematoma.
e G Angle
  H Ramus
  I Coronoid process
  J Head
  K Body

3.28 Answer

The parotid glands are paired salivary glands lying over the ramus and angle of the mandible. They are roughly triangular in shape.

The edge of the parotid runs from the tragus of the ear in a line just below and parallel to the zygomatic arch to just overlap the posterior border of the masseter muscle. From here, it arches back towards the angle of the mandible, which it wraps around. From this point it runs up, behind the ramus of the mandible, over the mastoid process and curves forward around the inferior part of the auricle of the ear.

3.29 Answer

The parotid duct runs in the middle third of a line drawn from the intertragic notch to the middle of the philtrum. It can be rolled over the masseter muscle and empties saliva into the mouth at the papilla opposite the second maxillary molar tooth.

3.30 Answer

The layers of the scalp are (from superficial to deep):
- Skin
- Connective tissue (containing vessels and nerves)
- Aponeurosis of occipitalis/frontalis muscles (galea)
- Loose connective (areolar) tissue
- Pericranium

Remember: SCALP.

3.31 Answer

The scalp receives blood from the:
- Supraorbital and supratrochlear arteries anteriorly
- Occipital artery posteriorly
- Superficial temporal and posterior auricular arteries laterally

3.32 Answer

The standard incision for submandibular gland surgery is 3 cm below the angle of the mandible. This is to avoid damaging the marginal mandibular branch of the facial nerve.

3.33 Answer

The submandibular gland lies in the digastric or submandibular triangle.

3.34 Answer

The following nerves are at risk during submandibular gland surgery:
- Lingual nerve
- Nerve to mylohyoid
- Hypoglossal nerve
- Marginal mandibular nerve

3.35 Answer

The facial artery and vein run over or through the gland and are usually ligated during submandibular gland surgery.

3.36 Answer

a A Middle concha
   B Pituitary gland
   C Transverse sinus
   D Falx cerebri
   E Tongue
   F Inferior concha
   G Ethmoid sinus
   H Superior concha
   I Epiglottis
   J Oesophagus
   K Mandible
   L Trachea
   M Frontal sinus
   N Hyoid bone

b O Superior meatus – lies between the superior and middle conchae and communicates with the ethmoidal sinuses
   P Middle meatus – lies between the middle and inferior conchae and communicates with the frontal sinus through the frontonasal ducts and   also the maxillary and ethmoidal sinuses
   Q Inferior meatus – lies below the inferior concha and communicates with the eye via the nasolacrimal duct
   R Sphenoethmoidal recess – lies above the superior concha and communicates with the sphenoid sinus
   S Eustachian (auditory) tube – links the middle ear to the nasopharynx

3.37 Answer

The superior orbital fissure is a slit-like foramen allowing communication between the middle cranial fossa and the orbit.

From superior to inferior the structures passing through this foramen are:
- Lacrimal nerve
- Frontal nerve
- Superior ophthalmic *vein*
- Trochlear nerve
- Superior division of the oculomotor nerve
- Nasociliary nerve
- Inferior division of oculomotor nerve
- Abducens nerve
- Inferior ophthalmic *vein*

3.38 Answer

There are seven extraocular muscles: one eyelid elevator, four recti muscles and two oblique muscles.

The *levator palpebrae superioris* originates from the sphenoid, just above the tendinous ring and inserts into the tarsal plate of the upper eyelid. It elevates the upper eyelid. Its dual nerve supply is the oculomotor nerve and sympathetic fibres.

The *superior rectus, inferior rectus, medial rectus* and *lateral rectus* muscles originate from the tendinous ring (on the sphenoid) and attach into the sclera of the eye in positions corresponding to their name. Each muscle pulls the eye in the direction corresponding to its name, i.e. the medial rectus pulls the eye medially and the lateral rectus pulls the eye laterally. However, the superior and inferior recti do not pull the eye directly up or down; they move slightly diagonally so that they slightly adduct the eye, i.e. the superior rectus pulls the eye up and medially, and the inferior rectus pulls the eye down and medially. This diagonal movement is corrected by the oblique muscles which also have a diagonal pull but in the lateral direction (abduction).

All of the recti are supplied by the oculomotor nerve apart from the lateral rectus, which is supplied by the abducens nerve (abducens abducts the eye).

The *superior oblique* originates just above the tendinous ring and passes obliquely forwards and medially, passing through a pulley (trochlear) and then attaching into the sclera posteriorly under the superior rectus. Its action is therefore to pull the eye inferiorly and laterally (down and out). The superior oblique is supplied by the trochlear nerve.

The *inferior oblique* originates from the anterior medial orbital floor and inserts into the posterior lateral sclera. It pulls the eye superiorly and laterally (up and out). The inferior oblique is supplied by the occulomotor nerve.

3.39 Answer

a A Trapezius
 B Sternocleidomastoid
 C Internal jugular vein
 D Carotid artery
 E Platysma
 F Investing layer of deep cervical fascia
 G Upper pole of thyroid gland
 H Spinal cord
 I Vagus nerve
 J Vertebral artery
 K Larynx
 L Pharynx (compressed)

b This cross-section is at the C4/C5 level.

Note: the carotid artery has not yet bifurcated (C4) and the vocal cords are just visible in the arytenoids cartilage.

3.40 Answer

The isthmus of the thyroid gland is at the level of the C7 vertebra.

3.41 Answer

The thyroid gland lies within its own covering of connective tissue – the thyroid fascia. This is within the pre-tracheal fascia, which surrounds the larynx and pharynx as well. The pre-tracheal and pre-vertebral fascia lie within the investing layer of deep cervical fascia.

3.42 Answer

The thyroid gland is supplied bilaterally by the superior and inferior thyroid arteries. The *superior thyroid artery* is a branch of the external carotid artery and supplies the upper pole. The *inferior thyroid artery* is a branch of the thyrocervical trunk and supplies the lower pole.

Three per cent of the population have a thyroidea ima artery, a midline artery from the brachiocephalic trunk that supplies the isthmus of the gland.

3.43 Answer

The *recurrent laryngeal nerve* runs close (usually posterior) to the inferior thyroid artery and is at risk of damage during ligation of this vessel. This nerve supplies all the muscles of the larynx except for the cricothyroid muscle. Damage to this nerve leads to paralysis of the ipsilateral vocal cord. Bilateral paralysis can lead to fatal occlusion of the airway.

The *external laryngeal nerve* runs close to the superior thyroid artery and is at risk of damage during ligation of this vessel. This nerve supplies the cricothyroid muscle and damage to it will manifest itself as a loss of timbre of the voice (monotone speech).

3.44 Answer

The thyroid gland is usually drained bilaterally by three veins:
- Superior thyroid vein
- Middle thyroid vein
- Inferior thyroid vein

The superior and middle veins empty into the internal jugular vein on that side. The inferior vein empties into the brachiocephalic vein.

4. Thorax

4.1 Answer

During inspiration, contraction of the intercostal muscles causes the middle of the ribs to rise upwards (bucket handle movement). This increases the transverse diameter of the chest. Movement of the ribs at the costovertebral joints also allows the anterior ends of the ribs to rise upwards (pump handle movement), thus increasing the anteroposterior diameter of the chest.

4.2 Answer

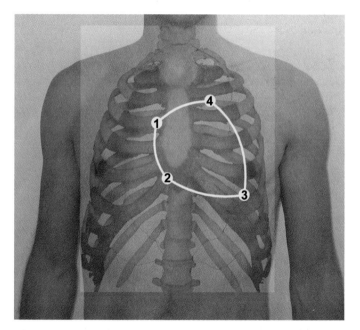

1. Lower border of the third costal cartilage at the right edge of the sternum
2. Lower border of the sixth costal cartilage at the right of the sternum
3. Fifth intercostal space in the mid-clavicular line
4. Second intercostal space, 2 cm to the left of the sternum

Draw a slightly curved line between these four points.

4.3 Answer

A chest drain is usually inserted into the fourth intercostal space in the anterior axillary line, just superior to the fifth rib (this avoids damaging the intercostal vessels that run in the upper part of the intercostal space, just inferior to the corresponding rib).

4.4   Answer

The drain passes through the:
- Skin
- Subcutaneous fat
- Deep fascia
- Serratus anterior muscle
- External intercostal muscle
- Internal intercostal muscle
- Innermost intercostal muscle
- Endothoracic fascia
- Parietal pleura

4.5   Answer

The intercostal neurovascular bundle runs in the plane between the internal intercostal and the innermost intercostal muscles in the order (from superior to inferior): vein, artery, nerve.

4.6   Answer

- Superior part – inferior thyroid arteries
- Middle part – oesophageal branches of the descending aorta
- Lower part – left gastric artery and inferior phrenic artery

4.7   Answer

The following structures cause a constriction in the oesophagus (from superior to inferior):
- Upper oesophageal sphincter
- Arch of aorta presses against the left lateral surface of the oesophagus
- Left main bronchus
- Diaphragmatic hiatus

4.8   Answer

The trachea bifurcates into the left and right main bronchi just below the level of the manubrio-sternal junction.

4.9   Answer

a   This is a contrast bronchogram.
b   A   Left main bronchus
    B   Right main bronchus
    C   Intermediate bronchus
    D   Trachea
    E   Right upper lobe bronchus

4.10   Answer

The right main bronchus is larger in diameter and forms less of an angle with the trachea than the left. For this reason a foreign body will tend to fall down the right side.

4.11 Answer

The *right lung* has three lobes: upper, middle and lower. The *left lung* has two lobes: upper and lower.

4.12 Answer

The *right lung* has 10 segments (three in the upper lobe, two in the middle lobe and five in the lower lobe). The *left lung* has nine or 10 segments, too (four or five in the upper lobe and five in the lower lobe).

4.13 Answer

A Jugular notch
B Manubrium
C Sternal angle (of Louis)
D Body
E Xiphoid process

4.14 Answer

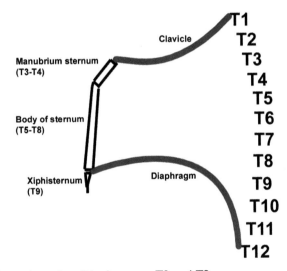

Sternal notch = Disc between T2 and T3
Manubrium = T3 to T4
Sternal angle (of Louis) = Disc between T4 and T5
Body of sternum = T5 to T8
Xiphisternum = T9

4.15 Answer

The right and left phrenic nerves are the motor supply to the diaphragm and arise from the C3, C4, C5 spinal levels.

Remember: C345 keeps the diaphragm alive.

4.16 Answer

The phrenic nerve provides sensation to the central part of the diaphragm, whilst the intercostal nerves T5-T11 and subcostal T12 provide sensation to the peripheral part.

4.17 Answer

Irritation of the central part of the diaphragm, e.g. from a bleeding splenic laceration, will be felt as referred pain in the dermatome of the phrenic nerve roots – C3,4,5, i.e. the shoulder. Irritation to the peripheral parts of the diaphragm is felt more locally.

4.18 Answer

The diaphragm is supplied by the following arteries:
- Superior and inferior phrenic arteries from the aorta
- Musculophrenic branches of the internal mammary artery
- A small contribution from the intercostal arteries to the peripheral parts of the diaphragm

4.19 Answer

The base of the breast is circular and extends from the lateral border of the sternum to the mid-axillary line, from the second rib down to the sixth rib.

4.20 Answer

The breast receives blood from four sources:
- Medially from branches of the internal mammary artery
- Laterally from branches of the lateral thoracic artery
- Deeply through perforators from the pectoralis major muscle (thoracoacromial trunk)
- Inferiorly/laterally from branches of the intercostal arteries

4.21 Answer

Sensation from the breast and nipple is carried by the fourth, fifth and sixth intercostal nerves.

4.22 Answer

a This is the medial aspect of the left lung. (The flat diaphragmatic base faces down whilst the thinner, sharper anterior border faces forwards.)
b D Aorta
   E Diaphragm
   F Heart
c G Phrenic nerve (anteriorly)
   H Vagus nerve (posteriorly)
d A Left main bronchus
   B Pulmonary artery
   C Pulmonary vein

4.23 Answer

Both lungs have an oblique fissure that extends from the spinous process of the T2 vertebra posteriorly to the sixth costal cartilage anteriorly. The right lung also has a horizontal fissure that lies under the fourth rib.

4.24 Answer

A Right auricle
B Ascending aorta
C Brachiocephalic trunk
D Arch of aorta
E Left anterior descending coronary artery (anterior interventricular artery)
F Right coronary artery
G Marginal artery
H Pulmonary trunk
I Left pulmonary artery
J Right ventricle
K Left ventricle
L Left carotid artery
M Left subclavian artery
N Superior vena cava
O Left main bronchus
P Inferior vena cava

4.25 Answer

a This is the superior aspect of a typical rib from the right side (this happens to be the eighth rib).

Note: The upper surface is blunt whereas the inferior border has a sharp edge, lateral to the costal groove. The neurovascular bundle runs behind this sharp edge.

b A Head
   B Neck
   C Tubercle
   D Shaft
   E Angle
c The costal cartilage attaches to this point.
d This is the tubercle of the rib and has a facet for articulation with the transverse process of the corresponding vertebra, e.g. if this is the fourth rib, the tubercle would articulate with the transverse process of the T4 vertebra.

4.26 Answer

a This is a right first rib. (Note how flat, how small and how tightly curved it is. The superior surface has the grooves.)
b A Head
   B Neck
   C Tubercle
   D Scalene tubercle

c The scalene tubercle (D) provides attachment for the scalenus anterior muscle.

d E Subclavian vein
F Subclavian artery

e The TI nerve root runs in direct contact with the shaft of the rib with the fibres of the C8 nerve root running above it. These nerve roots merge to form the inferior trunk of the brachial plexus.

f The C8 nerve root runs above the neck of the first rib and the TI nerve root runs below the neck. Both structures then pass above the shaft of the first rib (G).

g This is the head of the first rib and has a facet for articulation with the TI vertebra.

h The scalenus medius muscle attaches to the first rib at point H.

4.27 Answer

a This CT scan is at the level of the pulmonary arteries. It is therefore at the T5 vertebral level.

b A Ascending aorta
B Right pulmonary artery
C Descending aorta
D Body of T5 vertebra
E Superior vena cava
F Right main bronchus
G Left main bronchus
H Oesophagus
I Azygous vein
J Left scapula
K Left lower lobe pulmonary artery
L Left pulmonary vein
M Right pulmonary vein
N Sternum
O Left internal mammary artery

4.28 Answer

During subclavian vein cannulation the needle passes through the:
- Skin
- Subcutaneous fat
- Deep fascia
- Clavicular head of pectoralis major
- Clavipectoral fascia
- Subclavius
- Subclavian vein wall

4.29 Answer

At risk of damage are the:
- Subclavian artery
- Phrenic nerve
- Apex of lung
- Thoracic duct (on left side)

5. Abdomen and pelvis

5.1 Answer

Intraperitoneal:
- Stomach
- First part of the duodenum
- Jejunum
- Ileum
- Transverse colon
- Sigmoid colon
- Tail of pancreas
- Appendix
- Caecum
- Liver
- Gallbladder
- Uterus
- Ovaries

Retroperitoneal:
- Kidneys
- Ureters
- Adrenal glands
- Ascending and descending colon (including splenic and hepatic flexures)
- Most of the pancreas (except the tip of the tail which lies within the splenic peritoneum)
- second, third and fourth parts of duodenum
- Inferior vena cava
- Aorta
- Lymph nodes around the aorta
- Bladder
- Vagina

5.2 Answer

The jejunum is a darker red in colour, has a larger calibre lumen, a thicker wall, is more vascular, has longer vasa recta, fewer arcades, less fat in the mesentery and deeper plicae circulares.

5.3 Answer

The large bowel has sacculations/haustrations, appendices epiploicae, tenia coli, a larger calibre lumen and a thicker wall.

5.4 Answer

There are three large openings in the diaphragm:

- At the level of T8 is the *caval opening* which allows passage of the inferior vena cava, the right phrenic nerve and lymphatic vessels
- At the level of T10 is the *oesophageal hiatus* which allows passage of the oesophagus, vagus nerve, left gastric vessels and lymphatics
- At the level of T12 is the *aortic hiatus* which allows passage of the aorta, azygous vein and thoracic duct. This is actually an aperture posterior to the diaphragm rather than a true hole within it

5.5 Answer

T8 = Xiphisternal junction
T10 = Seventh costal cartilage
T12 = Just above the transpyloric plane

5.6 Answer

a This is a retrograde pyelogram.

Note: The contrast catheters inserted through the urethra and bladder and into the ureters. This is not an intravenous urogram where the contrast is given via a vein and then filtered by the kidneys. The appearance of the kidneys and ureters is similar in both investigations.

b A Left ureter
  B Right ureter
  C Left renal pelvis
  D Right pelvi-ureteric junction
  E Calyx
  F Bladder
c This line is the psoas shadow and is caused by the psoas major muscle.
d The renal papillae project into the minor calyces causing this characteristic shape. The minor calyces unite to form two or three major calyces which then empty into the renal pelvis.

5.7 Answer

The ureters start at the renal pelvis – the most posterior of the renal hilum structures (vein, artery, pelvis, from anterior to posterior). They are retroperitoneal for their entire course and there is an equal length of ureter in the abdomen and in the pelvis. The *abdominal ureter* descends almost vertically downwards, anterior to the psoas major muscle, just overlapping the transverse processes of the lumbar vertebrae. Each ureter is crossed anteriorly by the gonadal artery and vein. The genitofemoral nerve passes behind the ureter (which explains the referred pain to the testes with ureteric calculi).

At the pelvic brim the ureters pass anterior to the external iliac artery (ureter, artery, vein from anterior to posterior). The *left ureter* has the apex of the sigmoid mesocolon as its anterior relation at the pelvic brim. The *pelvic ureter* then descends posteroinferiorly on the lateral pelvic wall, anterior to the branches of the internal iliac arteries, then curves anteromedially to enter the posterolateral surface of the bladder. Just before entering the bladder the ureter is

crossed anteriorly by the vas deferens in the male or the uterine
artery in the female.

5.8 Answer

- Superior: ureteric branches of the renal arteries
- Middle: branches of the gonadal arteries
- Inferior: branches of the common and external iliac arteries
- Pelvic: branches of the internal iliac and vesical arteries

5.9 Answer

- Pelvic-ureteric junction
- Ureter changes direction as it crosses the pelvic brim
- Crossing of the gonadal artery
- Oblique intramural course of the vesiculo-ureteric junction

5.10 Answer

a This is a kidney.
b B Left renal vein
   C Left renal artery
   D Aorta
   E Inferior vena cava
   F Left ureter
c The left suprarenal (adrenal) gland.
d From anterior to posterior:
- Renal vein
- Renal artery (or arteries)
- Renal pelvis and ureter
e The renal artery is passing in front of the renal vein. This is
therefore the posterior aspect of the left kidney.
f The kidneys lie in the paravertebral grooves at the levels of T12 to
L3. The hilum is at the transpyloric plane (the left is slightly higher).
They are retroperitoneal structures. Each kidney is surrounded by
a renal capsule of fibrous tissue, which is within the perinephric fat.
This fat is then surrounded by the renal fascia (Gerota's fascia).
g The posterior relations of the kidney are (four muscles):
- Diaphragm (superiorly)
- Quadratus lumborum (inferiorly)
- Psoas major (medially)
- Transverses abdominis (laterally)

5.11 Answer

a This is an abdominal aortogram.
b A Abdominal aorta
   B Right common iliac artery
   C Left common iliac artery
   D Left renal artery
   E Right renal artery
   F Splenic artery
   G Common hepatic artery
   H Superior mesenteric artery

I   Gastroduodenal artery
J   Left hepatic artery
K   Right hepatic artery
L   Inferior mesenteric artery

5.12   Answer

The abdominal aorta starts in the midline as it passes through the diaphragm at the level of T12 (1 inch above the transpyloric plane). It runs in the retroperitoneum, in front of the lumbar vertebrae then bifurcates slightly to the left of the midline at the level of L4 (the supracrestal plane) into the common iliac arteries.

5.13   Answer

The *midline branches* are the:
• Coeliac axis – L1 upper
• Superior mesenteric artery – L1 lower
• Inferior mesenteric artery – L3
The *paired branches* are the:
• Inferior phrenic arteries – T12
• Suprarenal arteries – L1 (sometimes these arise from the renal arteries)
• Renal arteries – between L1 and L2
• Gonadal arteries – L2
• Four paired lumbar arteries – L1, 2, 3 and 4

5.14   Answer

The transpyloric plane is the line halfway between the suprasternal notch and the superior pubic symphesis.

At this level there is the:
• Termination of the spinal cord
• Aorta gives off the superior mesenteric artery
• Lateral border of the rectus abdominis muscle crosses the costal margin
• Duodenojejunal junction
• L1 vertebra
• Neck of the pancreas
• Hila of the kidneys
• Formation of the portal vein

5.15   Answer

• Skin
• Subcutaneous fat
• Scarpa's fascia
• Linea alba
• Transversalis fascia
• Extraperitoneal fat
• Peritoneum

5.16 Answer

The aponeuroses of three muscles make up the rectus sheath: external oblique, internal oblique and transversus abdominis.

The rectus sheath has an anterior layer and a posterior layer. The superior three-quarters are different from the inferior quarter. The arcuate line demarcates the boundary between the two regions and lies 3 cm below the umbilicus.

Superiorly, the anterior layer is made up of the aponeurosis of the external oblique muscle and half of the internal oblique muscle. The posterior layer is made up of the aponeurosis of the other half of the internal oblique and the transversus abdominis muscles.

Inferiorly, all three muscle layers pass anteriorly. There is no posterior layer of sheath. The rectus abdominis muscle is in direct contact with the transversalis fascia.

5.17 Answer

This is an opening between the greater sac (the main peritoneal cavity) and the lesser sac (omental bursa).

5.18 Answer

- Anterior: the free edge of the lesser omentum containing the portal triad
- Posterior: the inferior vena cava covered in peritoneum
- Superior: caudate process of the liver
- Inferior: the first part of the duodenum covered in peritoneum

5.19 Answer

This is Pringle's manoeuvre.

5.20 Answer

The portal triad is compressed between the finger and thumb as it runs through the free edge of the lesser omentum: portal vein posteriorly, common bile duct on the right and hepatic artery on the left.

5.21 Answer

This manoeuvre can be used temporarily to stop the inflow of blood to the liver, eg to control the haemorrhage from a liver laceration.

5.22 Answer

The hepatorenal space (Morrison's pouch) is the most dependent area when supine. This is continuous with the right paracolic gutter and lies between the right kidney and the liver.

5.23 Answer

This is the most likely site for a collection of blood or pus to pool, leading to the formation of an abscess.

5.24 Answer

The inguinal canal is a cleft-like space, entirely between the layers of the abdominal wall. It is approximately 2 cm wide and runs above and parallel to the inguinal ligament. It extends from the deep ring (opening in the transversalis fascia) to the superficial ring (opening in the external oblique).

5.25 Answer

- Anterior: external oblique aponeurosis, reinforced laterally by internal oblique
- Posterior: transversalis fascia, reinforced medially by the conjoint tendon
- Inferior: the in-rolled edge of external oblique (i.e. the inguinal ligament)
- Superior: the lower edge of internal oblique and transversus abdominis

5.26 Answer

Running through the inguinal canal is the:
- Ilioinguinal nerve
- Spermatic cord (males)
- Round ligament (females)

5.27 Answer

The mid-inguinal point is half way between the anterior superior iliac spine and the pubic symphysis. This corresponds to the femoral artery pulse.

5.28 Answer

The midpoint of the inguinal ligament is halfway between the anterior superior iliac spine and the pubic tubercle. This corresponds to the deep inguinal ring. The mid-inguinal point is therefore medial to the midpoint of the inguinal ligament.

5.29 Answer

a This is an endoscopic retrograde cholangiopancreatogram (ERCP).
b A Ampulla of Vater
  B Pancreatic duct
  C Common bile duct
  D Fundus of gallbladder
  E Body of gallbladder
  F Neck of gallbladder
  G Cystic duct
  H Common hepatic duct
  I Left hepatic duct
  J Right hepatic duct

5.30 Answer

The common bile duct enters the posteromedial wall of the second part of the duodenum at the ampulla of Vater.

5.31 Answer

The sphincter of Oddi regulates the flow of bile and is under neural and hormonal control.

5.32 Answer

As the testes descend through the inguinal canal with the vas deferens and neurovascular structures, they take a contribution from the abdominal wall layers, which become fascial coverings.

The layers of the scrotum and cord are:
- Skin from abdominal skin
- Dartos muscle from abdominal Camper's fascia
- Dartos fascia from abdominal Scarpa's fascia
- External spermatic fascia from external oblique muscle
- Cremaster muscle from internal oblique muscle
- Internal spermatic fascia from transversalis fascia
- Tunica vaginalis of the testes (and obliterated processus vaginalis of the cord) from the abdominal peritoneum

Note: The transversus abdominis muscle does not contribute a layer to the cord or scrotum.

5.33 Answer

The contents of the spermatic cord are:
- Three arteries: testicular, cremasteric and artery of the vas deferens
- Three nerves: sympathetic, parasympathetic and genitofemoral nerve
- Three other structures: vas deferens, pampaniform plexus of veins and lymphatic vessels

5.34 Answer

The testicular veins are formed by the merging pampaniform plexus of veins which serve an important role in thermoregulation of the testes. The left testicular vein drains into the left renal vein. The right testicular vein drains directly into the inferior vena cava.

5.35 Answer

Because of the more oblique entry of the right testicular vein into the inferior vena cava, it is less likely to allow backflow of blood. Varicocoeles are therefore more common on the left side.

5.36 Answer

The lumbar plexus is formed by the anterior rami of the L1 to L5 spinal nerves. The plexus lies within the psoas major muscle.

5.37 Answer

The nerves of the lumbar plexus are divided according to how they exit the psoas major muscle:

Lateral
- Iliohypogastric nerve (L1)
- Ilioinguinal nerve (L1)
- Lateral cutaneous nerve of thigh (L2, L3)
- Femoral nerve (L2, L3, L4)

Anterior
- Genitofemoral nerve (L1, L2)

Medial
- Obturator nerve (L2, L3, L4)
- Lumbosacral trunk (L4, L5)

5.38 Answer

a   This is a CT scan of the abdomen at the level of L1 – the transpyloric plane.
b   A   Right kidney
    B   Left kidney
    C   Liver
    D   Stomach
    E   Spleen
    F   Body of L1 vertebra
    G   Pancreas
    H   Aorta
    I   Left renal vein
    J   Inferior vena cava
    K   Superior mesenteric vein
    L   Right rectus abdominis muscle
    M   Second part of duodenum
    N   Right psoas major muscle
    O   Right quadratus lumborum muscle
    P   Superior mesenteric artery

5.39 Answer

The stomach is a foregut structure and is therefore supplied by branches of the coeliac trunk. The lesser curve of the stomach is supplied by the right and left gastric arteries. The greater curve of the stomach is supplied by the right and left gastroepiploic arteries. The fundus is supplied by the short gastric arteries.

The coeliac trunk divides into three branches: hepatic, splenic and left gastric arteries. The splenic artery gives off the short gastric and the left gastroepiploic arteries. The hepatic artery gives off the right gastric and the gastroduodenal arteries. The gastroduodenal artery divides into the superior pancreaticoduodenal and right gastroepiploic arteries.

5.40 Answer

The veins of the stomach correspond to the arteries:
- Lesser curve of the stomach is drained by the right and left gastric veins
- Greater curve of the stomach is drained by the right and left gastroepiploic   veins
- Fundus is drained by the short gastric veins
- Short gastric and left gastroepiploic veins drain into the splenic vein
- Right gastroepiploic veins drain into the superior mesenteric vein
- Right gastric vein drains directly into the portal vein – the portal vein is formed by the confluence of the superior mesenteric and splenic veins

Note: The inferior mesenteric vein drains into the splenic vein.

5.41 Answer

The portosystemic anastomoses occur at the:
- Oesophagus
- Umbilicus
- Rectum
- Retroperitoneum

5.42 Answer

The superior mesenteric artery provides the blood supply to the embryological midgut structures.
    It arises from the abdominal aorta at the transpyloric plane (L1) behind the body of the pancreas. It then passes anterior to the uncinate process of the pancreas and third part of the duodenum, i.e. the pancreas and duodenum are wrapped around the superior mesenteric artery and vein.

The branches of the superior mesenteric artery are the:
- Inferior pancreaticoduodenal
- Middle colic
- Right colic
- Several jejunal and ileal arteries
- Ileocolic

5.43 Answer

The colon is a midgut structure from the ileocaecal junction to the splenic flexure and is supplied by the superior mesenteric artery. The remaining portion of the colon (from the splenic flexure to the anus) is a hindgut structure and is therefore supplied by the inferior mesenteric artery.

The superior mesenteric artery gives the following branches to the colon:
- Ileocolic
- Right colic
- Middle colic

The *inferior mesenteric artery* gives the following branches to the colon:

- Left colic
- Several sigmoid arteries
- Superior rectal

These named arteries do not directly enter the colon but anastomose with each other to form the marginal artery (of Drummond) that runs alongside the colon, providing a collateral circulation.

5.44 Answer

This is a double contrast enema (barium and air).

A Caecum
B Appendix
C Ascending colon
D Transverse colon
E Descending colon
F Sigmoid colon
G Rectum
H Hepatic flexure
I Splenic flexure

5.45 Answer

The ascending colon, descending colon, hepatic and splenic flexures, and rectum are all retroperitoneal.

5.46 Answer

Calot's triangle is formed by the cystic duct laterally, the common hepatic duct medially and the edge of the liver superiorly. It contains the right hepatic artery, cystic artery and the right branch of the portal vein.

5.47 Answer

a This is the inferior surface of the liver. The anterior surface is towards the top of the picture.

b A Right lobe
B Left lobe
C Quadrate lobe
D Caudate lobe
c E Gallbladder
F Inferior vena cava
G Falciform ligament
H Ligamentum teres
I Common bile duct
J Portal vein
K Hepatic artery

5.48 Answer

   a This is the inferior surface of the spleen.
   b The spleen lies on the left side of the upper abdomen.

Note: This is a trick question to catch out those candidates who think this organ is a kidney – easily done in the stress of the exam situation when handed a shrivelled up specimen!

   c A Left kidney (renal impression)
       B Splenic flexure of colon (colonic impression)
       C Fundus/greater curve of stomach (gastric impression)
       D Tail of pancreas

Note: The superior/anterior surface usually has several notches to help you identify the gastric area.

   d The spleen is not normally palpable. It must be at least double its normal size before its border is palpable under the costal margin.

5.49 Answer

   a This is a female pelvis.
   b A Symphysis pubis
       B Bladder
       C Uterus
       D Rectum
       E Urethra
       F Vagina
       G Cervix of uterus
       H Rectouterine pouch (of Douglas)
       I Anal canal
       J Sigmoid colon
       K Body of L5 vertebra
       L Body of S1 vertebra
       M Cauda equina
   c This is the pelvic inlet.

5.50 Answer

The urethra starts at the bladder neck as the pre-prostatic urethra, descends through the prostate gland as the prostatic urethra and exits the prostate gland to become the membranous urethra, which then enters the bulb of the penis to become the spongy urethra. The spongy urethra passes through the penis to end as the external urethral meatus.

5.51 Answer

The membranous urethra is surrounded by the external urethral sphincter. This muscle relaxes during the voiding of urine.

5.52 Answer

The prostatic urethra is the most dilatable part of the urethra.

5.53 Answer

The membranous urethra is the least dilatable part of the urethra.

5.54 Answer

a This is the right hip bone (or innominate bone).
b A Anterior superior iliac spine
  B Iliac crest
  C Anterior inferior iliac spine
  D Obturator foramen
  E Greater sciatic notch
  F Ischial spine
  G Ischial tuberosity
  H Pubic tubercle
  I Lesser sciatic notch
  J Iliac fossa
  K Arcuate line
  L Superior ramus of pubis
  M Inferior ramus of pubis
  N Iliac tuberosity
c X Ischium
  Y Pubis
  Z Ilium
d The ilium, ischium and pubis meet in the acetabulum (meaning vinegar cup) to form the Mercedes symbol.
e The inguinal ligament attaches between the anterior superior iliac spine (A) and the pubic tubercle (H).
f O and P are the articular surfaces of the hip bone.
  O is the articular surface for the sacroiliac joint. This area is covered with hyaline cartilage. It is a synovial joint (the strongest in the body).
  P is the articular surface for the pubic symphysis. This area is covered by a thin layer of hyaline cartilage that attaches to the opposite side through a thick fibrocartilaginous disc. This is a secondary cartilaginous joint.
g The iliacus originates from the hollow of the iliac fossa, passes beneath the inguinal ligament and inserts into the psoas tendon.

5.55 Answer

- Superior third of the rectum has peritoneum anteriorly and laterally
- Middle third of the rectum has peritoneum anteriorly only
- Lower third of the rectum has no peritoneum

5.56 Answer

a This cystoscopy is from a male patient.
b A External urethral orifice
  B Penile urethra mucosa
  C Intrabulbar fossa
  D Opening to membranous urethra
  E Membranous urethra mucosa

F Verumontanum (seminal colliculus)
G Urethral crest
H Opening of prostatic utricle
I Prostatic sinus
J Internal urethral orifice

c The prostatic utricle is a small blind ending sac, a vestigial structure that is the homologue of the uterus. On either side of this (not seen in the photograph) is the slit-like opening of the ejaculatory ducts.

d The prostatic ducts open into the prostatic sinuses on either side of the urethral crest.

6. Back

6.1 Answer

There are 31 pairs of spinal nerves:
- Eight cervical
- Twelve thoracic
- Five lumbar
- Five sacral
- One coccygeal

6.2 Answer

A line drawn between the iliac crests (supracrestal plane) will intersect the vertebral column at the level of L4. A lumbar puncture needle should be inserted into the space between the L4 and L5 vertebrae.

6.3 Answer

The needle will pass through the:
- Skin
- Subcutaneous fat
- Deep fascia
- Supraspinous ligament
- Interspinous ligament
- Ligamentum flavum
- Dura mater
- Arachnoid mater

Note: There are no muscles in the midline.

6.4 Answer

The cell bodies of the sympathetic nerve cells lie within the lateral horn of grey matter of the spinal cord from the level of T1 to L2. Preganglionic fibres pass from these cell bodies through the anterior roots of the spinal cord to enter the anterior rami of the spinal nerves of T1 to L2.

From here these fibres pass through white rami communicantes to enter the sympathetic chain, where they synapse with post-ganglionic fibres. These synapses occur at the level of the nerve root or the preganglionic fibres and may ascend or descend a level. Some fibres pass through the sympathetic chain without synapsing to form splanchnic nerves.

The post-ganglionic fibres then return to the anterior rami via the grey rami communicantes for distribution to the target organs.

The sympathetic chain receives white rami communicantes from the T1 to L2 levels only but the chain extends up and down the entire length of the spinal column.

There are three cervical ganglia: superior, middle and inferior. Usually the inferior cervical ganglion is fused with the first thoracic ganglion to form the larger stellate ganglion, which lies just above the neck of the first rib.

Note: The white rami communicantes contain unmyelinated fibres whereas the grey rami communicantes contain myelinated fibres.

6.5   Answer

a This is a typical *cervical* vertebra (C3 to C7 are typical cervical vertebrae. This happens to be the C3 vertebra.)

b Characteristics of the cervical vertebrae are:
  • Large, triangular vertebral foramen
  • Small transverse processes containing a foramen
  • Small and wide body
  • Short and bifid (C3 to C5) spinous process

c A Spinous process
   B Posterior tubercle of transverse process
   C Anterior tubercle of transverse process
   D Superior articular process
   E Transverse foramen
   F Vertebral foramen
   G Body

6.6   Answer

a This is the inferior aspect of a thoracic vertebra (this is T6).

b Characteristics of the thoracic vertebrae are:
  • Smaller, circular vertebral foramen
  • Heart-shaped body
  • Long transverse processes that angle posteriorly
  • Facets on the sides of the body for articulation with the ribs
  • Long spinous process that slopes inferiorly

c A Body
   B Pedicle
   C Transverse process (sloping posteriorly)
   D Spinous process (sloping inferiorly)
   E Lamina
   F Inferior articular process

Further reading

Abrahams *et al. McMinn's Colour Atlas of Human Anatomy*, 4th edn.

McMinn. *Last's Anatomy: Regional and Applied*, 9th edn.

Moore. *Clinically Orientated Anatomy*, 5th edn.

Overstall S, Cunnick G, Mokbel K. Get Through Intercollegiate MRCS Parts 1 and 2: MCQs and EMQs. London: Royal Society of Medicine Press, 2005.

www.anatomedia.com